Studies in Classification, Data Analysis, and Knowledge Organization

For further volumes:
http://www.springer.com/1431-8814

Akinori Okada · Tadashi Imaizumi
Hans-Hermann Bock · Wolfgang Gaul
Editors

Cooperation in Classification and Data Analysis

Proceedings of Two German-Japanese Workshops

Editors

Professor Dr. Akinori Okada
Graduate School of Management and
Information Sciences
Tama University
4-1-1 Hijirigaoka, Tama-shi
Tokyo 206-0022
Japan
okada@tama.ac.jp

Professor Tadashi Imaizumi
School of Management and
Information Sciences
Tama University
4-1-1 Hijirigaoka, Tama-shi
Tokyo 206-0022
Japan
imaizumi@tama.ac.jp

Professor Dr. Dr. h.c. Hans-Hermann Bock
Institute of Statistics
RWTH Aachen University
52056 Aachen
Germany
bock@stochastik.rwth-aachen.de

Professor Dr. Wolfgang Gaul
Institute of Decision Theory and
Management Science
Faculty of Economics
University of Karlsruhe (TH)
76128 Karlsruhe
Germany
wolfgang.gaul@wiwi.uni-karlsruhe.de

ISSN 1431-8814
ISBN 978-3-642-00667-8 e-ISBN 978-3-642-00668-5
DOI: 10.1007/978-3-642-00668-5
Springer Dordrecht Heidelberg London New York

Library of Congress Control Number: 2009926244

Cover design: SPi Publisher Services

Printed on acid-free paper

Springer is part of Springer Science+Business Media (www.springer.com)

Preface

This volume contains selected papers presented at two joint German–Japanese symposia on data analysis and related fields. The articles substantially extend and further develop material presented at the two symposia organized on the basis of longstanding and close relationships which have been cultivated in the last couple of decades between the two classification societies: the German Classification Society (Gesellschaft für Klassifikation e. V.) and the Japanese Classification Society. These symposia have been very helpful in exchanging ideas, views, and knowledge between the two societies and have served as a spring board for more extensive and closer co-operation between the societies as well as among their individual members.

The scientific program of the first Joint Japanese–German Symposium (Tokyo 2005) included 23 presentations; for the second Joint German–Japanese Symposium (Berlin 2006) 27 presentations were scheduled. This volume presents 21 peer refereed papers, which are grouped into three parts:

1. Part 1 Clustering and Visualization (eight papers)
2. Part 2 Methods in Fields (nine papers)
3. Part 3 Applications in Clustering and Visualization (four papers)

The concept of having a joint symposium of the two classification societies came from the talks with Hans-Hermann and Wolfgang when Akinori attended the 28th Annual Conference of the German Classification Society held in Dortmund in March 2004.

From the Japanese side, Tadashi was the main local organizer of the first symposium in Tokyo and took the responsibility of putting together this volume (working closely with Akinori). From the German side Wolfgang (in close contact with Hans-Hermann) applied for DFG support to enable the participation of German scientists in the first symposium in Tokyo and took the responsibility of organizing the second symposium in Berlin.

We are indebted very much to many sponsors (in alphabetical order):

ARK Information Systems Co., LTD.
Behaviormetric Society of Japan
Biometric Society of Japan, Esumi Co., LTD.
Deutsche Forschungsgemeinschaft (DFG)
Information and Mathematical Science Laboratory, Inc.
Insight Research Co., Ltd.
INTAGE Inc.
Japan Statistical Society
Japanese Society of Computational Statistics
MATHEMATICAL SYSTEMS Inc.
Nippon Research Center, Ltd.
Shinagawa Prince Hotel
SPSS Japan Inc.
SYSLABO Inc.
Tama University
Tokyu Agency Inc.

who helped in organizing and financing the symposia and the publication of the present volume. We hope that this volume will further accelerate the growth of the friendship between the two classification societies and will help to widen and deepen the exchange of knowledge in all aspects of data analysis and clustering as well as of related fields.

Finally we thank the referees for their comments and suggestions w.r.t. the submitted papers and all those who helped to make the symposia in Berlin and Tokyo scientifically important and successful meeting points for our colleagues.

Tokyo, Karlsruhe, and Aachen
October 2008

Akinori Okada
Tadashi Imaizumi
Hans-Hermann Bock
Wolfgang Gaul

Contents

Part I Classification and Visualization

Analyzing Symbolic Data: Problems, Methods, and Perspectives 3
H.-H. Bock

Constraining Shape and Size in Clustering 13
C. Borgelt and R. Kruse

Dissolution and Isolation Robustness of Fixed Point Clusters 27
C. Hennig

ADCLUS: A Data Model for the Comparison of Two-Mode Clustering Methods by Monte Carlo Simulation 41
M. Wiedenbeck and S. Krolak-Schwerdt

Density-Based Multidimensional Scaling 53
F. Rehm, F. Klawonn, and R. Kruse

Classification of Binary Data Using a Spherical Distribution 61
Y. Sato

Fuzzy Clustering Based Regression with Attribute Weights 71
M. Sato-Ilic

Polynomial Regression with Some Observations Falling Below a Threshold 81
H.-Y. Siew and Y. Baba

Part II Methods in Fields

Feedback Options for a Personal News Recommendation Tool 91
C. Bomhardt and W. Gaul

Classification in Marketing Science 99
S. Scholz and R. Wagner

Deriving a Statistical Model for the Prediction of Spiralling in BTA Deep-Hole-Drilling from a Physical Model 107
C. Weihs, N. Raabe, and O. Webber

Analyzing Protein–Protein Interaction with Variant Analysis 115
G. Ritter and M. Gallegos

Estimation for the Parameters in Geostatistics 123
D. Niu and T. Tarumi

Identifying Patients at Risk: Mining Dialysis Treatment Data 131
T. Knorr, L. Schmidt-Thieme, and C. Johner

Sequential Multiple Comparison Procedure for Finding a Changing Point in Dose Finding Test 141
H. Douke and T. Nakamura

Semi-supervised Clustering of Yeast Gene Expression Data 151
Schönhuth, I.G. Costa, and A. Schliep

Event Detection in Environmental Scanning: News from a Hospitality Industry Newsletter 161
R. Wagner, J. Ontrup, and S. Scholz

Part III Applications in Clustering and Visualization

External Asymmetric Multidimensional Scaling Study of Regional Closeness in Marriage Among Japanese Prefectures 171
A. Okada

Socioeconomic and Age Differences in Women's Cultural Consumption: Multidimensional Preference Analysis 179
M. Nakai

Analysis of Purchase Intentions at a Department Store by Three-Way Distance Model 189
A. Nakayama

Facet Analysis of the AsiaBarometer Survey: Well-being, Trust and Political Attitudes 197
K. Manabe

Author Index 205

Subject Index 207

Contributors

Y. Baba The Institute of Statistical Mathematics, 4-6-7 Minami Azabu, Minato-ku, Tokyo 106-8569, Japan

H.-H. Bock Institute of Statistics, RWTH Aachen University, 52056 Aachen, Germany, bock@stochastik.rwth-aachen.de

C. Bomhardt Institut für Entscheidungstheorie und Unternehmensforschung, Universität Karlsruhe (TH), 76128 Karlsruhe, Germany

C. Borgelt European Center for Soft Computing Edificio Científico-Tecnológico, c/Gonzalo Gutiérrez Quirós s/n, 33600 Mieres, Asturias, Spain, christian.borgelt@softcomputing.es

I. G. Costa Max-Planck-Institut für Molekulare Genetik, 14195 Berlin, Germany

H. Douke Department of Mathematical Sciences, Tokai University, Hiratsuka, Kanagawa, 259-1292 Japan

M. Gallegos Faculty of Computer Science and Mathematics, University of Passau, 94030 Passau, Germany

W. Gaul Institut für Entscheidungstheorie und Unternehmensforschung, Universität Karlsruhe (TH), 76128 Karlsruhe, Germany, wolfgang.gaul@wiwi.uni-karlsruhe.de

C. Hennig Department of Statistical Science, University College London, Gower St, London WC1E 6BT, UK, chrish@stats.ucl.ac.uk

C. Johner Calcucare GmbH, Kaiser-Joseph-Str. 274, 79098 Freiburg, Germany

F. Klawonn Department of Computer Science, University of Applied Sciences, Braunschweig/Wolfenbüttel, Germany

T. Knorr Institut für Informatik, Universität Freiburg, 79110 Freiburg, Germany, till.knorr@googlemail.com

and

Calcucare GmbH, Kaiser-Joseph-Str. 274, 79098 Freiburg, Germany

S. Krolak-Schwerdt Department of Psychology, Saarland University, Germany, s.krolak@mx.uni-saarland.de

R. Kruse Department of Knowledge Processing and Language Engineering, Otto-von-Guericke-University of Magdeburg, Universitätsplatz 2, 39106 Magdeburg, Germany

K. Manabe School of Sociology, Kwansei Gakuin University, 1-1-155 Uegahara, Nishinomiya 662-0811 Japan, kazufumi.manabe@nifty.com

M. Nakai Department of Social Sciences, College of Social Sciences, Ritsumeikan University, 56-1 Toji-in Kitamachi, Kita-ku, Kyoto 603-8577, Japan, mnakai@ss.ritsumei.ac.jp

T. Nakamura Tokai University, Hiratsuka, Kanagawa, 259-1292 Japan, tomonaka@keyaki.cc.u-tokai.ac.jp

A. Nakayama Faculty of Economics, Nagasaki University, 4-2-1 Katafuchi, Nagasaki, Japan 850-8506, atsuho@nagasaki-u.ac.jp

D. Niu Graduate School of Natural Science and Technology, Okayama University, 1-1, Naka 3-chome, Tsushima, Okayama 700-8530, Japan, ndhui@f7.ems.okayama-u.ac.jp

A. Okada Graduate School of Management and Information Sciences, Tama University, 4-1-1 Hijirigaoka, Tama-shi, Tokyo, 206-0022 Japan, okada@tama.ac.jp

J. Ontrup Neuroinformatics Group, Bielefeld University, 33615 Bielefeld, Germany

N. Raabe Chair of Computational Statistics, University of Dortmund, Germany

F. Rehm Institute of Flight Guidance, German Aerospace Center, Braunschweig, Germany, frank.rehm@dlr.de

G. Ritter Faculty of Computer Science and Mathematics, University of Passau, 94030 Passau, Germany, ritter@fim.uni-passau.de

Y. Sato Division of Computer Science, Graduate School of Information Science and Technology, Hokkaido University, Kita 14, Nishi 9, Kita-ku, Sapporo, 060-0814 Japan, ysato@main.ist.hokudai.ac.jp

M. Sato-Ilic Department of Risk Engineering, Faculty of Systems and Information Engineering, University of Tsukuba, Tennodai 1-1-1, Tsukuba, Ibaraki 305-8573, Japan, mika@sk.tsukuba.ac.jp

A. Schliep Max-Planck-Institut für Molekulare Genetik, 14195 Berlin, Germany, asa86@cs.sfu.ca

L. Schmidt-Thieme Institut für Informatik, Universität Freiburg, 79110 Freiburg, Germany

S. Scholz Business Administration and Marketing, Bielefeld University, 33615 Bielefeld, Germany

A. Schönhuth ZAIK, Universität zu Köln, 50931 Köln, Germany

H.-Y. Siew Department of Statistical Science, The Graduate University for Advanced Studies, 4-6-7 Minami Azabu, Minato-ku, Tokyo 106-8569, Japan

R. Wagner DMCC Dialog Marketing Competence Center, University of Kassel, 34125 Kassel, Germany, rwagner@wirtschaft.uni-kassel.de

O. Webber Department of Machining Technology, University of Dortmund, Germany

C. Weihs Chair of Computational Statistics, University of Dortmund, Germany, claus.weihs@T-Online.de

M. Wiedenbeck Centre for Survey Research and Methodology, Mannheim, Germany

T. Tarumi Admission Centre, Okayama University, 1-1, Naka 3-chome, Tsushima, Okayama 700-8530, Japan

Part I
Classification and Visualization

Analyzing Symbolic Data: Problems, Methods, and Perspectives

H.-H. Bock

Abstract Classical data analysis considers data vectors with real-valued or categorical components. In contrast, *Symbolic Data Analysis (SDA)* deals with data vectors whose components are intervals, sets of categories, or even frequency distributions. SDA generalizes common methods of multivariate statistics to the case of symbolic data tables. This paper presents a brief survey on basic problems and methods of this fast-developing branch of data analysis. As an alternative to the current more or less heuristic approaches, we propose a new probabilistic approach in this context. Our presentation concentrates on visualization, dissimilarities, and partition-type clustering for symbolic data.

1 Introduction

Classical data analysis considers single-valued variables such that, for n objects and p variables, each entry x_{kj} of the data matrix $X = (x_{kj})_{n\times p}$ is a real number (quantitative type) or a category (qualitative type). The term *symbolic data* relates to more general scenarios where the entries x_{kj} may be either an interval $x_{kj} = [a_{kj}, b_{kj}] \in I\!R$, a set of categories $x_{kj} = \{\alpha, \beta, \ldots\}$, or even a frequency distribution. This situation can be met, e.g., when X summarizes the properties of n student groups ($\hat{=}$ rows) and the k-th row vector

$$x_k = ([20, 25], \{\text{math, phys, chem}\}, (D: 0.6,\ NL: 0.2,\ JAP: 0.2))'$$

indicates that the k-th group comprises students between 20 and 30 years, studying mathematics, physics, or chemistry, and originating from Germany (60%), Belgium

H.-H. Bock
Institute of Statistics, RWTH Aachen University, 52056 Aachen, Germany,
E-mail: bock@stochastik.rwth-aachen.de

A. Okada et al. (eds.), *Cooperation in Classification and Data Analysis*,
Studies in Classification, Data Analysis, and Knowledge Organizations.
DOI: 10.1007/978-3-642-00668-5_1,

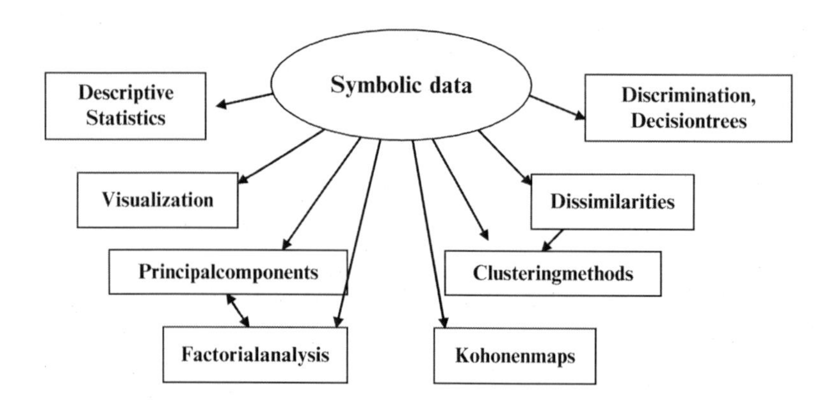

(20%), and Japan (20%). As illustrated by the diagram, numerous methods and algorithms have been developed for the analysis of symbolic data and many of them were integrated into the software package SODAS (see, e.g., Bock and Diday 2000, Diday and Noirhomme 2008).

Typically these methods are more or less sophisticated generalizations of classical data analysis tools with the proviso that thus far most approaches are based on empirical, heuristical or informatics-related argumentations, but not so much on probability models. This paper presents a brief survey on the basic elements of symbolic data analysis in the domain of clustering, and proposes a probabilistic approach as an alternative.

Thereby, we will concentrate on the case of p interval variables such that $X = ([a_{kj}, b_{kj}])$ is a $n \times p$ matrix of univariate intervals. Then the k-th object is described by the vector of intervals $x_k = ([a_{k1}, b_{k1}], \ldots, [a_{kp}, b_{kp}])'$ and can be represented as a p-dimensional (hyper-)rectangle $Q_k = [a_{k1}, b_{k1}] \times \cdots \times [a_{kp}, b_{kp}] = [a_k, b_k]$ in $I\!R^p$ where $a_k = (a_{k1}, \ldots, a_{kp})'$ and $b_k = (b_{k1}, \ldots, b_{kp})'$ are the "lower" and "upper" vertex of Q_k, respectively.

2 Visualization Tools: Zoom Stars and Principal Component Analysis

Since p-dimensional rectangles cannot be directly visualized on the screen, the following visualization tools have been developed:

(a) Zoom stars:

Each object k is visualized by a symmetric "star" where p coordinate lines originate from a common center (with an angle of $360/p$ between two adjacent lines) and the j-th interval $[a_{kj}, b_{kj}]$ is displayed on the j-th coordinate line. Visual comparison of the zoom stars of all n objects may reveal the underlying structure and peculiarities of the data set.

(b) Symbolic principal component analysis:

The n rectangles $Q_1, \ldots, Q_n$ are projected onto a common, two-dimensional hyperplane H in $\mathbb{R}^p$. H is the two-dimensional principal component hyperplane that is obtained either from the n midpoints $m_k = (a_k + b_k)/2$ of the rectangles Q_k (center method) or from the $n \cdot 2^p$ vertices of the rectangles Q_k (vertices method). Since SDA adheres to the idea that the output of a method should be formulated in the same terms as the input data (here: rectangles), an object k is *not* represented by the projection Q_k' (polytope) of the Q_k onto H (i.e., on the screen), but by the smallest rectangle R_k that contains the projection Q_k' and whose sides are parallel to the first two principal component axes.

(c) Kohonen maps:

Kohonen maps display the interval data X in a two-dimensional rectangular (or hexagonal) lattice of "prototypes" (neurons) and can thereby reveal the overall structure of the set of objects. The most simple approach proceeds by applying the classical self-organizing Kohonen algorithm (see, e.g. Bock 1999) to the $2p$-dimensional "data" points $y_k := (a_k', b_k')'$ that represent the lower and upper vertices of the rectangles Q_k. There exist several other options for this approach (see, e.g., Bock 2003, 2008, El Golli et al. 2004).

3 Dissimilarity Between Data Rectangles

Dissimilarities between objects are the basis for data analysis methods such as multidimensional scaling, classification, and clustering. Therefore, a basic step in SDA consists in the definition of a suitable dissimilarity measure for symbolic data vectors. In the interval data case, this amounts to define formally the dissimilarity of two p-dimensional rectangles $Q = [a, b] = [a_1, b_1] \times \cdots \times [a_p, b_p]$ and $R = [u, v] = [u_1, v_1] \times \cdots \times [u_p, v_p]$ from $\mathbb{R}^p$. We concentrate here on measures that evaluate the "geographical" position of the rectangles in $\mathbb{R}^p$. But there are also measures that consider, additionally, the shapes (elongated, flat, cubic,...) and overlap of Q and R (see, e.g., Bock and Diday 2000, Chap. 8).

(a) Vertex-type distance d_v

The vertex-type distance between Q and R is defined by

$$d_v(Q, R) := \left\| \begin{pmatrix} a \\ b \end{pmatrix} - \begin{pmatrix} u \\ v \end{pmatrix} \right\|^2 \tag{1}$$

$$= ||a - u||^2 + ||b - v||^2 = \sum_{j=1}^{p} ((a_j - u_j)^2 + (b_j - v_j)^2).$$

Symbolic methods based on this dissimilarity are often similar to classical data analysis methods applied to the $2p$-dimensional vectors $y_k = (a'_k, b'_k)'$, $k = 1, \ldots, n$.

(b) Hausdorff distance

Given an arbitrary metric d on $I\!R^p$ (here: the Euclidean metric), the *Hausdorff distance* between two sets $Q, R \subset I\!R^p$ is given by

$$d_H(Q, R) := \max\{\delta(Q, R), \delta(R, Q)\},$$

where

$$\delta(Q, R) := \max_{\beta \in R} \min_{\alpha \in Q} d(\alpha, \beta)$$

is the minimum distance of a point β from the rectangle Q, maximized over all $\beta \in R$. The distance d_H is often used in classical calculus and geometrical probability when characterizing the convergence of a set Q to a fixed set R. For intervals in R^1 it reduces to

$$d_H([a, b], [u, v]) = \max\{|a - u|, |b - v|\} =: D_1([a, b], [u, v]). \qquad (2)$$

Whereas the computation of d_H is computationally prohibitive for two arbitrary sets in R^p if $p \geq 2$, there exists a recurrence algorithm with moderate computational complexity for the case of two p-dimensional intervals (Bock 2005). Nevertheless, for computational convenience, most SDA methods use one of the following heuristic modifications.

(c) Hausdorff-type L_q-distance:

Here the p one-dimensional Hausdorff distances between the coordinate-wise intervals $[a_j, b_j]$ and $[u_j, v_j]$ are combined into one single measure by the Minkowski-type formula (with a $q \geq 1$, typically $q = 1, 2, \infty$):

$$d_p^{(q)} := \left(\sum_{j=1}^{p} D_1^q([a_j, u_j], [b_j, v_j]) \right)^{1/q} = \left(\sum_{j=1}^{p} \max\{|a_j - u_j|^q, |b_j - v_j|^q\} \right)^{1/q}.$$

Various authors have adapted this measure to the case of missing values or logical dependencies among the coordinate intervals.

4 Average Intervals and Class Prototypes: Centrocubes

When considering a class (cluster) $C = \{1, \ldots, n\}$ of objects described by n intervals $Q_1 = [a_1, b_1], \ldots, Q_n = [a_n, b_n]$ we are often interested in a "class prototype" and must therefore define some "average" of the n intervals $Q_1, \ldots, Q_n$ or, more generally, the expectation of a "random set". This question has been investigated by many authors in mathematics, geometric probability and data analysis.

(a) Centrocubes as optimum class representatives

A basic approach for defining the average of n intervals $Q_1, \dots, Q_n$ in a class $\mathcal{G}$ of intervals in $I\!R^p$ starts from a dissimilarity measure d between two intervals and defines the "class prototype" or "centrocube" as the interval $G = G(C) = [u, v] = ([u_1, v_1], \dots, [u_p, v_p])$ in $\mathcal{G}$ that minimizes the deviation measure

$$g(C, G) := \sum_{k \in C} d(Q_k, G) \to \min_{G \in \mathcal{G}}. \quad (3)$$

Typically there will be no explicit solution for this problem, but there are special cases where the solution can easily be obtained (Bock 2005):

(α) If d is the vertex-type distance (1), the centrocube is given by $G(C) = [\hat{a}_C, \hat{b}_C]$ where $\hat{a}_C, \hat{b}_C$ are the arithmetic means of the n lower and upper vertices of the n data intervals, respectively.

(β) If d is the Hausdorff-type L_1 distance, the solution is given by median values: For each coordinate j, let $m_{kj} = (a_{kj} + b_{kj})/2$ be the midpoint and $\lambda_{kj} = (b_{kj} - a_{kj})/2$ the midrange of the data interval Q_k such that $Q_k = [a_{kj}], b_{kj}] = [m_{kj} - \lambda_{kj}, m_{kj} + \lambda_{kj}]$. Then the centrocube $G(C) = ([u_1, v_1], \dots, [u_p, v_p])$ is given by the coordinate intervals $[u_j, v_j] = [\mu_j - \lambda_j, \mu_j + \lambda_j]$ with median midpoints $\mu_j := median(m_{1j}, \dots, m_{nj})$ and median midranges $\lambda_j := median(\lambda_{1j}, \dots, \lambda_{nj})$ (Chavent and Lechevallier 2002).

(b) Approaches starting from geometric probability

Whereas the previous definitions of "prototype intervals" were more or less based on heuristic grounds, there exist various attempts to define set averages in the framework of geometrical probability (where in our case, the empirical probability measure must be considered, assigning mass $1/n$ to each object). This requires first the exact definition of a p-dimensional "random (closed) set Q" (not necessarily an interval!) in the framework of measure theory (see, e.g., Mathéron 1975). Then we may look for a definition of the "expectation" $E[Q]$ of the random set with the hope that many useful properties of the classical concept of the "expectation for a random variable" can be retained or are at least valid in an extended sense. A wealth of definitions and properties are presented and surveyed, e.g., in Kruse (1987), Molchanov (1997), Baddeley and Molchanov (1997, 1998), and Nordhoff (2003).

5 Partitioning Clustering Methods

Clustering objects into a system $\mathcal{C} = (C_1, \ldots, C_m)$ of classes is a common tool for analyzing the structure of a (classical or symbolic) data set. Whereas classes may overlap and the number m of classes may be unknown in the general case, we concentrate here on methods that start from n p-dimensional data hypercubes $Q_1, \ldots, Q_n$ and look for an optimum partition $\mathcal{C}$ of the underlying set of objects $\mathcal{O} := \{1, \ldots, n\}$ with a known number m of classes. Since clusters must be illustrated and interpreted in practice, we also should look for a system of m optimum class prototypes $z_1, \ldots, z_m$, i.e., a system $\mathcal{Z} = (z_1, \ldots, z_m)$ of m rectangles in $\mathbb{R}^p$.

A wealth of classical optimality criteria can be adapted to this symbolic case, for example,

$$g_1(\mathcal{C}, \mathcal{Z}) := \sum_{i=1}^{m} \sum_{k \in C_i} d(Q_k, z_i) \to \min_{\mathcal{C}, \mathcal{Z}}, \tag{4}$$

$$g_2(\mathcal{C}) := \sum_{i=1}^{m} \sum_{k \in C_i} \sum_{l \in C_i} d(Q_k, Q_l) \to \min_{\mathcal{C}}. \tag{5}$$

The criterion g_2, a sum of cluster heterogeneities, must be minimized with respect to all m-partitions $\mathcal{C}$. This is a combinatorial optimization problem for which can be solved by linear or combinatorial optimization methods if the number of objects n is not too large (see, e.g., Hansen and Jaumard 1997).

In contrast, the criterion g_1 depends on two "parameters" $\mathcal{C}$ and $\mathcal{Z}$ and must be minimized with respect to both of them, typically under the constraint that the prototype hypercubes z_i belong to a prespecified system $\mathcal{G}$ of hypercubes. In order to approximate an optimum pair $(\mathcal{C}, \mathcal{Z})$ we can invoke the generalized k-means approach that decreases g_1 step by step: Starting from an arbitrary (e.g., a random) system of initial class prototypes $\mathcal{Z}^{(0)}$ we proceed by:

(a) First determining the minimum-distance partition $\mathcal{C}^{(0)}$ of $\mathcal{O}$ with classes

$$C_i^{(0)} := \left\{ k \in \mathcal{O} \mid d\left(Q_k, z_i^{(0)}\right) = min_j \; d\left(Q_k, z_j^{(0)}\right) \right\} \quad \text{for } i = 1, \ldots, m$$

(with some device for solving ties and avoiding empty classes)

(b) Then calculating, for each class $C_i^{(0)}$, a new class prototype (centrocube) $z_i^{(1)}$ according to

$$\sum_{k \in C_i^{(0)}} d(Q_k, z_i) \to min_{z_i \in \mathcal{G}} \tag{6}$$

(c) And then iterating (a) and (b) until stationarity

The results from Sect. 4 show that this k-means-type algorithm can be easily implemented if d is the vertex-type distance (1) (see, e.g., Carvalho et al. 2005) or the

Hausdorff-type L_1 distance since in these cases there exists an explicit formula for the centrocubes (if $\mathcal{G}$ is the set of all rectangles in $\mathbb{R}^p$).

Another case where a direct solution is possible is provided by the case when $\mathcal{G} = \{Q_1, \ldots, Q_n\}$ is the set of *observed* data rectangles, or minimization in (6) is constrained to all elements of the i-th class such that (6) amounts to representing the i-th class by its "most typical" *data* rectangle. Writing $z_i = Q_{k_i}$ with k_i an element from class i, we may here rewrite (4) in the form:

$$g_3(\mathcal{C}, \mathcal{G}) := \sum_{i=1}^{m} \sum_{k \in C_i} d(Q_k, Q_{k_i}) \rightarrow \min_{\mathcal{C}, k_i \in C_i \text{ for all } i} \tag{7}$$

and the most typical object, the *medoid interval* of $C_i^{(0)}$, is easily determined by (6).

As in the classical case, it is obvious that these k-means algorithms may converge to a local minimum that is not a global minimum of the clustering criterion. Also we may apply other heuristic minimization algorithms such as an exchange of single objects between classes, simulated annealing, genetic algorithms, etc.

6 A Parametric Probabilistic Approach for Clustering Interval Data

It is astonishing that most adherents and users of SDA, whenever they generalize a classical data analysis method that is based on a probability model, do not use a probabilistic argumentation as well, but only some heuristic or informatics-type considerations. In contrast, we will briefly sketch here how parametric probability models can be introduced into this symbolic context.

Classical statistics will consider the n data intervals as independent samples of an underlying "random" rectangle. A probability model for a random rectangle $Q = [U, V] = [U_1, V_1] \times \cdots \times [U_p, V_p]$ in $\mathbb{R}^p$ has to specify the joint distribution of the interval boundaries $U_1, \ldots, V_p$ or, more conveniently, of the interval midpoints $M_j := (U_j + V_j)/2$ and the midranges $L_j := (V_j - U_j)/2$ for $j = 1, \ldots, p$. For example, we may design a *normal-gamma model* as follows:

(1) The vector of midpoints $M = (M_1, \ldots, M_p)'$ is independent from the vector of midranges $L = (L_1, \ldots, L_p)'$.
(2) The vector M has a p-dimensional normal distribution $\mathcal{N}_p(\mu, \Sigma)$ with expectation $\mu \in \mathbb{R}^p$ and a positive definite covariance matrix Σ (with $\Sigma = \sigma^2 \cdot I_p$ in the "spherical" case); let $h_1(m; \mu, \Sigma)$ the density of M.
(3) The midranges $L_1, \ldots, L_p$ are all independent and for all j L_j has a gamma distribution $Gamma(\alpha_j, \beta_j)$ with a coordinate-specific size parameter $\alpha_j > 0$ and a shape parameter $\beta_j > 0$.

Note that $\Gamma(\alpha_j, \beta_j)$ has the density $q(s; \alpha_j, \beta_j) := (\beta_j^{\alpha_j} / \Gamma(\alpha_j)) \cdot s^{\alpha_j - 1} exp(-\beta_j s)$ for $s > 0$ with $E[L_j] = \alpha_j / \beta_j$ and $Var(L_j) = \alpha_j / \beta_j^2$. Let us denote by $h_2(\ell; a, b) := \Pi_j q(\ell_j; \alpha_j, \beta_j)$ the density of $L = (L_1, \ldots, L_p)'$ and by

$\mathcal{I}(\mu, \Sigma, a, b)$ the resulting distribution for (M, L) with the parameter vectors $a := (\alpha_1, \ldots, \alpha_p)'$, $b := \beta_1, \ldots, \beta_p)' \in R_+^p$. Then (M, L) has the joint density $f(m, \ell; \vartheta) := h_1(m; \mu, \Sigma) \cdot h_2(\ell; a, b)$ where ϑ comprises the unknown parameters.

In this model a reasonable definition of a "prototype interval" for Q might be the p-dimensional interval

$$\mathcal{M} := [\, E[M] - E[L],\; E[M] + E[L]\,]$$

that is centered at $E[M] = \mu$ with midrange vector $E[L] = (\alpha_1/\beta_1, \ldots, \alpha_p/\beta_p)'$.

In order to estimate the unknown parameters $E[M] = \mu, \Sigma, \sigma^2, E[L], \alpha_j, \beta_j$ from n independent samples $Q_1, \ldots, Q_n$ of Q, we may use the maximum likelihood method. If $m_1, \ldots, m_n \in \mathbb{R}^p$ are the observed midpoints and $\ell_1, \ldots, \ell_n$ the observed midrange vectors of $Q_1, \ldots, Q_n$ with $\ell_k = (\ell_{k1}, \ldots, \ell_{kp})'$, the m.l. estimates are given by

$$\hat{\mu} = \bar{m} := \frac{1}{n} \sum_{k=1}^{n} m_k, \tag{8}$$

$$\hat{\Sigma} = \frac{1}{n} \sum_{k=1}^{n} (m_k - \bar{m})(m_k - \bar{m})' \quad \text{in general,}$$

$$\hat{\sigma}^2 = \frac{1}{np} \sum_{k=1}^{n} ||m_k - \bar{m}||^2 \qquad \text{in the spherical model,} \tag{9}$$

whereas the estimates $\hat{\alpha_j}$ and $\hat{\beta_j}$ are the solutions of the m.l. equations

$$\ln \hat{\alpha}_j - \psi(\hat{\alpha}_j) = ln\left(\bar{\ell}_j / \widetilde{\ell}_j\right) \qquad \hat{\beta}_j = \hat{\alpha}_j / \bar{\ell}_j, \tag{10}$$

where $\bar{\ell}_j := \frac{1}{n} \sum_{k=1}^{n} \ell_{kj}$ and $\widetilde{\ell}_j := (\Pi_k \ell_{kj})^{1/(n)}$ are the arithmetic and geometric mean of $\ell_{1j}, \ldots, \ell_{nj}$, respectively, and $\psi(z) := \Gamma'(z)/\Gamma(z)$ is the digamma function (for details see, e.g., Johnson et al. 1994, p. 360, or Kotz et al. 2006, p. 2625).

If we replace the Gamma distribution for the midranges L_j by a uniform distribution in an interval $[0, \Delta_j]$ from $\mathbb{R}_+^p$, the m.l. estimate of the boundary Δ_j is given by $\hat{\Delta}_j := \max_k\{\ell_{kj}\}$.

Clustering models can be formulated along the same lines, either as a "fixed-partition" model, a "random partition" model, or a mixture model (see Bock 1996a,b,c). As an example let us consider the following *symbolic fixed-partition model* for the random intervals $Q_1, \ldots, Q_n$:

(1) There exists an unknown partition $\mathcal{C} = (C_1, \ldots, C_m)$ of $\mathcal{O}$ with a known number of classes $C_i \subset \mathcal{O}$.
(2) There exist m sets of class-specific parameters (μ_i, a_i, b_i) (for $i = 1, \ldots, m$) with $\mu_i \in \mathbb{R}^p$ and $a_i = (\alpha_{i1}, \ldots, \alpha_{ip})', b_i = (\beta_{i1}, \ldots, \beta_{ip})'$ in $\mathbb{R}_+^p$, and $\sigma^2 > 0$.

(3) All intervals Q_k from the same class C_i have the same distribution:

$$Q_k \sim \mathcal{I}(\mu_i, \sigma^2 I_p, a_i, b_i) \qquad \text{for } k \in C_i.$$

Maximizing the likelihood with respect to the system ϑ of all parameters and to the m-partition $\mathcal{C}$ yields the *m.l. clustering criterion*:

$$g_4(\mathcal{C}, \vartheta) := \Pi_{i=1}^{m} \Pi_{k \in C_i} h_1(m_k; \mu_i, \sigma^2 I_p) \cdot h_2(\ell_k; a_i, b_i) \rightarrow min_{\mathcal{C},\vartheta}. \tag{11}$$

A solution, in particular an optimum m-partition, can be approximated by the classical *k-means-type algorithm*:

Starting from an initial partition $\mathcal{C} = (C_1, \ldots, C_m)$:

(a) We determine, in each class C_i, the m.l. estimates $\hat{\mu}_i, \hat{a}_i, \hat{b}_i, \hat{\sigma}^2$ from (8) to (10).

(b) Then build m new classes $C_1^{(1)}, \ldots, C_m^{(1)}$ by assigning each object k ($\hat{=}$ data interval Q_k) to the class with maximum likelihood such that for $i = 1, \ldots, m$:

$$C_i^{(1)} := \{k \in \mathcal{O} \,|\, f(m_k, \ell_k; \hat{\mu}_i, \hat{\sigma}^2, \hat{a}_i, \hat{b}_i) = \max_{j=1,\ldots,m} f(m_k, \ell_k; \hat{\mu}_j, \hat{\sigma}^2, \hat{a}_j, \hat{b}_j)\}.$$

(c) And iterating (a) and (b) until stationarity.

7 Final Remarks

In this paper we have presented some data analysis methods for symbolic data where data are intervals in $\mathbb{R}^p$. Typically, such methods result from heuristic adaptations of classical methods, with new types of dissimilarities and class prototypes. However, in our last section, we have shown that we may also use the classical approach of statistics: Starting from a (parametric) statistical model for "random intervals", the well-known estimation and clustering approach can be used and leads to a new k-means-type clustering algorithm for interval data. Analogous extensions and algorithms can be formulated in the case of the "random-partition" model or the mixture model from classical model-based cluster analysis. It might be an interesting perspective to compile a list of such models and methods and to compare their results in practical cases with the results from the "standard" symbolic data analysis.

References

Baddeley, A.J., and Molchanov, I.S. (1997): On the expected measure of a random set. In: D. Jeulin (ed.): *Advances in theory and applications of random sets.* World Scientific, Singapore, 3–20.

Baddeley, A.J., and Molchanov, I.S. (1998): Averaging of random sets based on their distance functions. *Journal of Mathematical Imaging and Vision* 8, 79–92.

Bock, H.-H. (1996a): Probability models and hypotheses testing in partitioning cluster analysis. In: Ph. Arabie, L. Hubert, and G. De Soete (Eds.): *Clustering and classification.* World Science, River Edge, NJ, 1996, 377–453.

Bock, H.-H. (1996b): Probabilistic models in cluster analysis. *Computational Statistics and Data Analysis* 23, 5–28.

Bock H.-H. (1996c): Probabilistic models in partitional cluster analysis. In: A. Ferligoj and A. Kramberger (Eds.): *Developments in data analysis.* FDV, Metodoloski zvezki, 12, Ljubljana, Slovenia, 1996, 3–25.

Bock, H.-H. (1999): Clustering and neural network approaches. In: W. Gaul, and H. Locarek-Junge (Eds): *Classification in the information age.* Studies in Classification, Data Analysis, and Knowledge Organization. Springer, Heidelberg, 1999, 42–57.

Bock, H.-H. (2003): Clustering methods and Kohonen maps for symbolic data. *Journal of the Japanese Society of Computational Statistics* 15.2, 217–229.

Bock, H.-H. (2005): Optimization in symbolic data analysis: dissimilarities, class centers, and clustering. In: D. Baier, R. Decker, and L. Schmidt-Thieme (eds.): *Data analsis and decision support.* Studies in Classification, Data Analysis, and Knowledge Organization. Springer, Heidelberg, 3–10.

Bock, H.-H. (2008): Visualizing symbolic data by Kohonen maps. In: E. Diday, and M. Noirhomme (Eds.): *Symbolic data analysis and the SODAS software.* Wiley, Chichester, 2008, 205–234.

Bock, H.-H., and Diday, E. (2000): *Analysis of symbolic data. Exploratory methods for extracting statistical information from complex data.* Studies in Classification, Data Analysis, and Knowledge Organization. Springer, Heidelberg.

Chavent, M., and Lechevallier, Y. (2002): Dynamical clustering of interval data: Optimization of an adequacy criterion based on Hausdorff distance. In: K. Jajuga, A. Sokolowski, H.-H. Bock (eds.): Classification, clustering, and data analysis. Springer, Berlin - Heidelberg, 2002, 53–60.

De Carvalho, F., Brito, B., and Bock, H.-H. (2005): Dynamic clustering for interval data based on L_2 distance. *Computational Statistics* 21, 231–250.

Diday, E., and Noirhomme, M. (Eds.) (2008): *Symbolic data analysis and the SODAS software.* Wiley, Chichester.

El Golli, A., Conan-Guez, B., and Rossi, F. (2004): A self-organizing map for dissimilarity data. In: D. Banks, L. House, F.R. McMorris, P. Arabie, and W. Gaul (Eds.): *Classification, clustering, and data mining applications.* Studies in Classification, Data Analysis, and Knowledge Organization. Springer, Heidelberg, 61–68.

Hansen, P., and Jaumard, B. (1997): Cluster analysis and mathematical programming. *Mathematical Programming* 79, 191–215.

Johnson, N.L., Kotz, S., and Balakrishnan, N (1994): *Continuous univariate distributions,* Vol. 1. Wiley, New York.

Kotz, S., Balakrishnan, N., Read, C.B., and Vidakovic, B. (2006): *Encyclopedia of statistical sciences,* Vol. 4. Wiley, New York.

Kruse, R. (1987): On the variance of random sets. *Journal of Mathematical Analysis and Applications* 122(2), 469–473.

Mathéron, G. (1975): *Random sets and integral geometry.* Wiley, New York.

Molchanov, I. (1997): Statistical problems for random sets. In: J. Goutsias (Ed.): *Random sets: theory and applications.* Springer, Berlin, 27–45.

Nordhoff, O. (2003): *Expectation of random intervals (in German: Erwartungswerte zufälliger Quader).* Diploma thesis. Institute of Statistics, RWTH Aachen University, 2003.

Constraining Shape and Size in Clustering

C. Borgelt and R. Kruse

Abstract Several of the more sophisticated fuzzy clustering algorithms, like the Gustafson–Kessel algorithm and the fuzzy maximum likelihood estimation (FMLE) algorithm, offer the possibility to induce clusters of ellipsoidal shape and differing sizes. The same holds for the expectation maximization (EM) algorithm for a mixture of Gaussian distributions. However, these additional degrees of freedom can reduce the robustness of the algorithms, thus sometimes rendering their application problematic, since results are unstable. In this paper we suggest methods to introduce shape and size constraints that handle this problem effectively.

1 Introduction

Prototype-based clustering methods, like fuzzy clustering variants (Bezdek 1981, Bezdek et al. 1999, Höppner et al. 1999), expectation maximization (EM) (Dempster et al. 1977) of a mixture of Gaussian distributions (Everitt and Hand 1981), or learning vector quantization (Kohonen 1986, 1995), often employ a distance function to measure the similarity of two data points. If this distance function is the *Euclidean distance*, all clusters are (hyper-) spherical. However, more sophisticated approaches rely on a cluster-specific *Mahalanobis distance*, making it possible to find clusters of (hyper-)ellipsoidal shape. In addition, they relax the restriction (as it is present, for example, in the fuzzy c-means algorithm) that all clusters have the same size (Keller and Klawonn 2003). Unfortunately, these additional degrees of freedom often reduce the robustness of the clustering algorithm, thus sometimes rendering their application problematic, since results can become unstable.

In this paper we consider how shape and size parameters of a cluster can be controlled, that is, can be constrained in such a way that extreme cases are ruled out

C. Borgelt(✉)
European Center for Soft Computing Edificio Científico-Tecnológico, c/Gonzalo Gutiérrez Quirós s/n, 33600 Mieres, Asturias, Spain, E-mail: christian.borgelt@softcomputing.es

A. Okada et al. (eds.), *Cooperation in Classification and Data Analysis*,
Studies in Classification, Data Analysis, and Knowledge Organizations.
DOI: 10.1007/978-3-642-00668-5_2,

and/or a bias (of user-specified strength) against extreme cases is introduced, which effectively improves robustness. The basic idea of constraining shape is the same as that of Tikhonov regularization for linear optimization problems (Tikhonov and Arsenin 1977, Engl et al. 1996), while size and weight constraints can be based on a bias towards equality as it is well-known from Laplace correction or Bayesian approaches to probability estimation, which introduce a similar bias towards a uniform distribution.

This paper is organized as follows: in Sects. 2 and 3 we briefly review some basics of mixture models (Gaussian mixture models in particular) and the expectation maximization algorithm as well as fuzzy clustering, which will be the method to which we apply our approach. In Sect. 4 we discuss our procedures to constrain shape, size, and weight parameters in clustering. In Sect. 5 we present experimental results on well-known data sets and finally, in Sect. 6, we draw conclusions from our discussion.

2 Mixture Models and the EM Algorithm

In a mixture model it is assumed that a data set $\mathcal{X} = \{\mathbf{x}_j \mid j = 1, \ldots, n\}$ has been sampled from a population of c clusters (Everitt and Hand 1981). Each cluster is characterized by a probability distribution, specified as a prior probability and a conditional probability density function (cpdf). The data generation process can be imagined like this: first a cluster i, $i \in \{1, \ldots, c\}$, is chosen, indicating the cpdf to be used, and then a datum is sampled from this cpdf. Consequently the probability of a data point $\mathbf{x}$ can be computed as

$$p_{\mathbf{X}}(\mathbf{x}; \Theta) = \sum_{i=1}^{c} p_C(i; \Theta_i) \cdot f_{\mathbf{X}|C}(\mathbf{x}|i; \Theta_i),$$

where C is a random variable describing the cluster i chosen in the first step, $\mathbf{X}$ is a random vector describing the attribute values of the data point, and $\Theta = \{\Theta_1, \ldots, \Theta_c\}$ with each Θ_i containing the parameters for one cluster (that is, its prior probability $\theta_i = p_C(i; \Theta_i)$ and the parameters of its cpdf).

Assuming that the data points were drawn independently from the same distribution (that is, that the distributions of their underlying random vectors $\mathbf{X}_j$ are identical), we can compute the probability of a given data set $\mathcal{X}$ as

$$P(\mathcal{X}; \Theta) = \prod_{j=1}^{n} \sum_{i=1}^{c} p_{C_j}(i; \Theta_i) \cdot f_{\mathbf{X}_j|C_j}(\mathbf{x}_j|i; \Theta_i),$$

Note, however, that we do not know which value the random variable C_j, which indicates the cluster, has for each example case $\mathbf{x_j}$. Fortunately, though, given the data point, we can compute the posterior probability that a data point $\mathbf{x}$ has been sampled from the cpdf of the i-th cluster using Bayes' rule:

$$p_{C|\mathbf{X}}(i|\mathbf{x};\Theta) = \frac{p_C(i;\Theta_i) \cdot f_{\mathbf{X}|C}(\mathbf{x}|i;\Theta_i)}{f_{\mathbf{X}}(\mathbf{x};\Theta)}$$
$$= \frac{p_C(i;\Theta_i) \cdot f_{\mathbf{X}|C}(\mathbf{x}|i;\Theta_i)}{\sum_{k=1}^{c} p_C(k;\Theta_k) \cdot f_{\mathbf{X}|C}(\mathbf{x}|k;\Theta_k)}.$$

This posterior probability may be used to complete the data set w.r.t. the cluster, namely by splitting each datum $\mathbf{x}_j$ into c data points, one for each cluster, which are weighted with the posterior probability $p_{C_j|\mathbf{X}_j}(i|\mathbf{x}_j;\Theta)$. This idea is used in the well-known expectation maximization (EM) algorithm (Dempster et al. 1977), which consists in alternatingly computing these posterior probabilities, using them as case weights, and estimating the cluster parameters from the completed data set by maximum likelihood estimation.

For clustering metric data it is usually assumed that the cpdf of each cluster is an m-variate normal distribution (so-called *Gaussian mixture model*) (Everitt and Hand 1981, Bilmes 1997). That is, the cpdf is assumed to be

$$f_{\mathbf{X}|C}(\mathbf{x}|i;\Theta_i) = N(\mathbf{x};\boldsymbol{\mu}_i,\Sigma_i)$$
$$= \frac{1}{\sqrt{(2\pi)^m|\Sigma_i|}} \exp\left(-\frac{1}{2}(\mathbf{x}-\boldsymbol{\mu}_i)^\top \Sigma_i^{-1}(\mathbf{x}-\boldsymbol{\mu}_i)\right),$$

where $\boldsymbol{\mu}_i$ is the mean vector and Σ_i the covariance matrix of the normal distribution, $i = 1,\dots,c$, and m is the number of dimensions of the data space. In this case the maximum likelihood estimation formulae are

$$\theta_i = \frac{1}{n}\sum_{j=1}^{n} p_{C|\mathbf{X}_j}(i|\mathbf{x}_j;\Theta)$$

for the prior probability θ_i,

$$\boldsymbol{\mu}_i = \frac{\sum_{j=1}^{n} p_{C|\mathbf{X}_j}(i|\mathbf{x}_j;\Theta) \cdot \mathbf{x}_j}{\sum_{j=1}^{n} p_{C|\mathbf{X}_j}(i|\mathbf{x}_j;\Theta)}$$

for the mean vector $\boldsymbol{\mu}_i$, and

$$\Sigma_i = \frac{\sum_{j=1}^{n} p_{C|\mathbf{X}_j}(i|\mathbf{x}_j;\Theta) \cdot (\mathbf{x}_j-\boldsymbol{\mu}_i)(\mathbf{x}_j-\boldsymbol{\mu}_i)^\top}{\sum_{j=1}^{n} p_{C|\mathbf{X}_j}(i|\mathbf{x}_j;\Theta)}$$

for the covariance matrix Σ_i of the i-th cluster, $i = 1,\dots,c$.

3 Fuzzy Clustering

While most classical clustering algorithms assign each datum to exactly one cluster, thus forming a crisp partition of the given data, fuzzy clustering allows for *degrees of membership*, to which a datum belongs to different clusters (Bezdek 1981, Bezdek et al. 1999, Höppner et al. 1999). Fuzzy clustering algorithms are usually objective function based: they determine an optimal (fuzzy) partition of a given data set $\mathbf{X} = \{\mathbf{x}_j \mid j = 1, \ldots, n\}$ into c clusters by minimizing an objective function

$$J(\mathbf{X}, \mathbf{U}, \mathbf{C}) = \sum_{i=1}^{c} \sum_{j=1}^{n} u_{ij}^{w} \, d_{ij}^{2}$$

subject to the constraints

$$\forall i; 1 \leq i \leq c: \sum_{j=1}^{n} u_{ij} > 0 \quad \text{and} \quad \forall j; 1 \leq j \leq n: \sum_{i=1}^{c} u_{ij} = 1,$$

where $u_{ij} \in [0, 1]$ is the membership degree of datum $\mathbf{x}_j$ to cluster i and d_{ij} is the distance between datum $\mathbf{x}_j$ and cluster i. The $c \times n$ matrix $\mathbf{U} = (u_{ij})$ is called the *fuzzy partition matrix* and $\mathbf{C}$ describes the set of clusters by stating location parameters (i.e., the cluster center) and maybe size and shape parameters for each cluster. The parameter w, $w > 1$, is called the *fuzzifier* or *weighting exponent*. It determines the "fuzziness" of the classification: with higher values for w the boundaries between the clusters become softer, with lower values they get harder. Usually $w = 2$ is chosen. Hard or crisp clustering results in the limit for $w \to 1$. However, a hard assignment may also be determined from a fuzzy result by assigning each data point to the cluster to which it has the highest degree of membership.

The left constraint guarantees that no cluster is empty (has no data points assigned to it) and the right constraint ensures that each datum has the same total influence by requiring that its membership degrees sum to 1. Due to the second constraint this approach is called *probabilistic fuzzy clustering*, because with it the membership degrees for a datum at least formally resemble the probabilities of its being a member of the corresponding clusters. The partitioning property of a probabilistic clustering algorithm, which "distributes" the weight of a datum to the different clusters, is due to this constraint.

Unfortunately, the objective function J cannot be minimized directly. The most popular approach to handle this problem is an iterative algorithm, which alternatingly optimizes membership degrees and cluster parameters (Bezdek 1981, Bezdek et al. 1999, Höppner et al. 1999). That is, first the membership degrees are optimized for fixed cluster parameters, then the cluster parameters are optimized for fixed membership degrees. The main advantage of this scheme is that in each of the two steps the optimum can be computed directly. By iterating the two steps the joint optimum is approached. (Although, of course, it cannot be guaranteed that the global optimum will be reached – one may get stuck in a local minimum of the objective function J.)

The update formulae are derived by simply setting the derivative of the objective function J w.r.t. the parameters to optimize equal to zero (necessary condition for a minimum). Independent of the chosen distance measure we thus obtain the following update formula for the membership degrees (Bezdek 1981, Bezdek et al. 1999, Höppner et al. 1999):

$$u_{ij} = \frac{d_{ij}^{-\frac{2}{w-1}}}{\sum_{k=1}^{c} d_{kj}^{-\frac{2}{w-1}}}. \tag{1}$$

Thus the membership degrees represent the relative inverse squared distances of a data point to the different cluster centers, which is a very intuitive result.

The update formulae for the cluster parameters, however, depend on what parameters are used to describe a cluster (location, shape, size) and on the distance measure. Therefore a general update formula cannot be given. Here we briefly review the three most common cases: The best-known fuzzy clustering algorithm is the fuzzy c-means algorithm, which is a straightforward generalization of the classical crisp c-means algorithm. It uses only cluster centers for the cluster prototypes and relies on the *Euclidean distance*, i.e.,

$$d_{ij}^2 = (\mathbf{x}_j - \boldsymbol{\mu}_i)^\top (\mathbf{x}_j - \boldsymbol{\mu}_i),$$

where $\boldsymbol{\mu}_i$ is the center of the i-th cluster. Consequently it is restricted to finding spherical clusters of equal size. The resulting update rule is

$$\boldsymbol{\mu}_i = \frac{\sum_{j=1}^{n} u_{ij}^w \mathbf{x}_j}{\sum_{j=1}^{n} u_{ij}^w}, \tag{2}$$

that is, the new cluster center is the weighted mean of the data points assigned to it, which is again a fairly intuitive result.

The Gustafson–Kessel algorithm (Gustafson and Kessel 1979), on the other hand, does not use a global distance measure, but employs a cluster-specific *Mahalanobis distance*:

$$d_{ij}^2 = (\mathbf{x}_j - \boldsymbol{\mu}_i)^\top \Sigma_i^{-1} (\mathbf{x}_j - \boldsymbol{\mu}_i).$$

Here $\boldsymbol{\mu}_i$ is the cluster center and Σ_i is a cluster-specific covariance matrix with determinant 1. It describes the shape of the cluster, thus allowing for ellipsoidal clusters of equal size (due to the fixed determinant, which is a measure of the cluster size). This distance measure leads to the same update rule (2) for the clusters centers, while the covariance matrices are updated as

$$\Sigma_i = \frac{\Sigma_i^*}{\sqrt[m]{|\Sigma_i^*|}}, \quad \text{where} \quad \Sigma_i^* = \frac{\sum_{j=1}^{n} u_{ij}^w (\mathbf{x}_j - \boldsymbol{\mu}_i)(\mathbf{x}_j - \boldsymbol{\mu}_i)^\top}{\sum_{j=1}^{n} u_{ij}^w} \tag{3}$$

and m is the number of dimensions of the data space. Σ_i^* is called the *fuzzy covariance matrix*, which is normalized to determinant 1 to meet the above-mentioned constraint. Compared to standard statistical estimation procedures, this is also a fairly intuitive result. It should be noted that the restriction to clusters of equal size may be relaxed by simply allowing general covariance matrices. However, depending on the characteristics of the data, this additional degree of freedom can deteriorate the robustness of the algorithm.

Finally, the fuzzy maximum likelihood estimation (FMLE) algorithm (Gath and Geva 1989) is based on the assumption that the data was sampled from a mixture of c multivariate normal distributions as in the statistical approach of mixture models (cf. Sect. 2). It uses a (squared) distance that is inversely proportional to the probability that a datum was generated by the normal distribution associated with a cluster and also incorporates the prior probability of the cluster. That is, the distance measure is

$$d_{ij}^2 = \left[\frac{\theta_i}{\sqrt{(2\pi)^m |\Sigma_i|}} \exp\left(-\frac{1}{2} (\mathbf{x}_j - \boldsymbol{\mu}_i)^\top \Sigma_i^{-1} (\mathbf{x}_j - \boldsymbol{\mu}_i) \right) \right]^{-1}.$$

Here θ_i is the prior probability of the cluster, $\boldsymbol{\mu}_i$ is the cluster center, Σ_i a cluster-specific covariance matrix, which in this case is not required to be normalized to determinant 1, and m the number of dimensions of the data space (cf. Sect. 2). For the FMLE algorithm the update rules are not derived from the objective function due to technical obstacles, but by comparing it to the expectation maximization (EM) algorithm for a mixture of normal distributions (cf. Sect. 2). This analogical reasoning leads to the same update rules for the cluster center and the cluster-specific covariance matrix as for the Gustafson–Kessel algorithm (Höppner et al. 1999), that is, (2) and (3) apply. The prior probability θ_i is, also in analogy to statistical estimation (cf. Sect. 2), computed as

$$\theta_i = \frac{1}{n} \sum_{j=1}^{n} u_{ij}^w. \tag{4}$$

Note that the difference to the EM algorithm consists in the different ways in which the membership degrees (1) and the posterior probabilities in the EM algorithm are computed and used in the estimation.

Since the high number of free parameters of the FMLE algorithm renders it unstable on certain data sets, it is usually recommended (Höppner et al. 1999) to initialize it with at least a few steps of the very robust fuzzy c-means algorithm. The same holds, though to a somewhat lesser degree, for the Gustafson–Kessel algorithm.

It is worth noting that of both the Gustafson–Kessel algorithm and the FMLE algorithm there exist so-called *axes-parallel* versions, which restrict the covariance matrices Σ_i to diagonal matrices and thus allow only axes-parallel ellipsoids (Klawonn and Kruse 1997). These constrained variants have certain advantages w.r.t. robustness and execution time (since the needed matrix inverses can be computed much faster than for full covariance matrices).

4 Constraining Cluster Parameters

The large number of parameters (mainly the elements of the covariance matrices) of the more flexible fuzzy and probabilistic clustering algorithms can render these algorithms less robust or even fairly unstable, compared to their simpler counterparts that only adapt the cluster centers (of spherical clusters). Common undesired results include very long and thin ellipsoids as well as clusters collapsing to a single data point. To counteract such undesired tendencies, we introduce shape and size constraints into the update scheme. The basic idea is to modify, in every update step, the parameters of a cluster in such a way that certain constraints are satisfied or at least that a noticeable tendency (of varying strength, as specified by a user) towards satisfying these constraints is introduced. In particular we consider constraining the (ellipsoidal) shape (by regularizing the covariance matrix) as well as constraining the (relative) size and the (relative) weight of a cluster.

4.1 Constraining Cluster Shapes

The shape of a cluster is represented by its covariance matrix Σ_i. Intuitively, Σ_i describes a general (hyper-)ellipsoidal shape, which can be obtained, for example, by computing the Cholesky decomposition or the eigenvalue decomposition of Σ_i and mapping the unit (hyper-)sphere with this decomposition.

Constraining the shape means to modify the covariance matrix, so that a certain relation of the lengths of the major axes of the represented (hyper-)ellipsoid is obtained or that at least a tendency towards this relation is introduced. Since the lengths of the major axes are the roots of the eigenvalues of the covariance matrix, constraining the axes relation means shifting the eigenvalues of Σ_i. Note that such a shift leaves the eigenvectors unchanged, that is, the orientation of the (hyper-)ellipsoid is preserved. Note also that such a shift of the eigenvalues is the basis of the well-known Tikhonov regularization for linear optimization problems (Tikhonov and Arsenin 1977, Engl et al. 1996), which inspired our approach. We suggest two methods:

Method 1: The covariance matrices Σ_i, $i = 1, \ldots, c$, of the clusters are adapted (after every standard update step) according to

$$\Sigma_i^{(\text{adap})} = \sigma_i^2 \cdot \frac{\mathbf{S}_i + h^2\mathbf{1}}{\sqrt[m]{|\mathbf{S}_i + h^2\mathbf{1}|}} = \sigma_i^2 \cdot \frac{\Sigma_i + \sigma_i^2 h^2\mathbf{1}}{\sqrt[m]{|\Sigma_i + \sigma_i^2 h^2\mathbf{1}|}},$$

where m is the dimension of the data space, $\mathbf{1}$ is a unit matrix, $\sigma_i^2 = \sqrt[m]{|\Sigma_i|}$ is the equivalent isotropic variance (equivalent in the sense that it leads to the same (hyper-)volume, that is, $|\Sigma_i| = |\sigma_i^2\mathbf{1}|$), $\mathbf{S}_i = \sigma_i^{-2}\Sigma_i$ is the covariance matrix scaled to determinant 1, and h is the regularization parameter.

This modification of the covariance matrix shifts all eigenvalues by the value of $\sigma_i^2 h^2$ and then renormalizes the resulting matrix so that the determinant of the old covariance matrix is preserved (that is, the (hyper-)volume of the cluster is kept constant). This adaptation tends to equalize the lengths of the major axes of the represented (hyper-)ellipsoid and thus introduces a tendency towards (hyper-)spherical clusters. (Algebraically, it makes the matrix “less singular”, and thus “more regular”, which explains the name *regularization* for this modification.) This tendency of equalizing the axes lengths is the stronger, the greater the value of h. In the limit, for $h \to \infty$, the clusters are forced to be exactly spherical. On the other hand, if $h = 0$, the cluster shape is left unchanged.

Method 2: The first method always changes the length ratios of the major axes and thus introduces a general tendency towards (hyper-)spherical clusters, which may be undesirable for properly shaped clusters. Therefore in this (second) method a limit r, $r > 1$, for the length ratio of the longest to the shortest major axis is used and only if this limit is exceeded, the eigenvalues are shifted in such a way that the limit is satisfied.

Formally: let λ_k, $k = 1, \ldots m$, be the eigenvalues of the covariance matrix Σ_i. Compute (after every standard update step)

$$h^2 = \begin{cases} 0, & \text{if } \dfrac{\max_{k=1}^m \lambda_k}{\min_{k=1}^m \lambda_k} \leq r^2, \\ \dfrac{\max_{k=1}^m \lambda_k - r^2 \min_{k=1}^m \lambda_k}{\sigma_i^2 (r^2 - 1)}, & \text{otherwise,} \end{cases}$$

and then execute Method 1 with this value of h^2.

4.2 Constraining Cluster Sizes

The size of a cluster can be described in different ways, for example, by the determinant of its covariance matrix Σ_i, which is a measure of the clusters squared (hyper-)volume, an equivalent isotropic variance σ_i^2 or an equivalent isotropic radius (standard deviation) σ_i (where the variance and the radius are equivalent in the sense that they lead to the same (hyper-)volume, see above). The latter two measures are defined as

$$\sigma_i^2 = \sqrt[m]{|\Sigma_i|} \qquad \text{and} \qquad \sigma_i = \sqrt{\sigma_i^2} = \sqrt[2m]{|\Sigma_i|}$$

and thus the (hyper-)volume of a cluster may also be written as $\sigma_i^m = \sqrt{|\Sigma_i|}$.

Constraining the (relative) cluster size means to ensure a certain relation between the cluster sizes or at least to introduce a tendency into this direction. We suggest three different versions of modifying cluster sizes, in each of which the measure that is used to describe the cluster size is specified by an exponent a of the equivalent isotropic radius σ_i. Special cases are

$$\begin{aligned} a &= 1 : \text{equivalent isotropic radius,} \\ a &= 2 : \text{equivalent isotropic variance,} \\ a &= m : \text{(hyper-)volume.} \end{aligned}$$

Method 1: The equivalent isotropic radii σ_i are adapted (after every standard update step) according to

$$\sigma_i^{(\text{adap})} = \sqrt[a]{s \cdot \frac{\sum_{k=1}^{c} \sigma_k^a}{\sum_{k=1}^{c} (\sigma_k^a + b)} \cdot (\sigma_i^a + b)}$$
$$= \sqrt[a]{s \cdot \frac{\sum_{k=1}^{c} \sigma_k^a}{cb + \sum_{k=1}^{c} \sigma_k^a} \cdot (\sigma_i^a + b)}.$$

That is, each cluster size is increased by the value of the parameter b and then the sizes are renormalized so that the sum of the cluster sizes is preserved. However, the parameter s may be used to scale the sum of the sizes up or down (by default $s = 1$). For $b \to \infty$ the cluster sizes are equalized completely, for $b = 0$ only the parameter s has an effect. This method is inspired by Laplace correction or Bayesian estimation with an uniformative prior (see below).

Method 2: This method, which is meant as a simplified and thus more efficient version of method 1, does not renormalize the sizes, so that the size sum is increased by cb. However, this missing renormalization may be mitigated to some degree by specifying a value of the scaling parameter s that is smaller than 1 and thus acts in a similar way as the renormalization. Formally, the equivalent isotropic radii σ_i are adapted (after every standard update step) according to

$$\sigma_i^{(\text{adap})} = \sqrt[a]{s \cdot (\sigma_i^a + b)}.$$

Method 3: The first two methods always change the relation of the cluster sizes and thus introduce a general tendency towards clusters of equal size, which may be undesirable for properly sized clusters. Therefore in this (third) method a limit r, $r > 1$, for the size ratio of the largest to the smallest cluster is used and only if this limit is exceeded, the sizes are changed in such a way that the limit is satisfied. To achieve this, b is set (after every standard update step) according to

$$b = \begin{cases} 0, & \text{if } \frac{\max_{k=1}^{c} \sigma_k^a}{\min_{k=1}^{c} \sigma_k^a} \leq r, \\ \frac{\max_{k=1}^{c} \sigma_k^a - r \min_{k=1}^{c} \sigma_k^a}{r - 1}, & \text{otherwise,} \end{cases}$$

and then Method 1 is executed with this value of b.

4.3 Constraining Cluster Weights

A cluster weight θ_i appears only in the mixture model approach and the FMLE algorithm, where it describes the prior probability of a cluster. For cluster weights we may use basically the same adaptation methods as for the cluster size, with the exception of the scaling parameter s (since the θ_i are probabilities, i.e., we must ensure $\sum_{i=1}^{c} \theta_i = 1$). Therefore we have the following two methods:

Method 1: The cluster weights θ_i are adapted (after every standard update step) according to

$$\theta_i^{(\text{adap})} = \frac{\sum_{k=1}^{c} \theta_k}{\sum_{k=1}^{c} (\theta_k + b)} \cdot (\theta_i + b) = \frac{\sum_{k=1}^{c} \theta_k}{cb + \sum_{k=1}^{c} \theta_k} \cdot (\theta_i + b),$$

where b is a parameter that is to be specified by a user. Note that this method is equivalent to a Laplace corrected estimation of the prior probabilities or a Bayesian estimation with an uninformative (uniform) prior.

Method 2: The value of the adaptation parameter b appearing in Method 1 is computed (after every standard update step) as

$$b = \begin{cases} 0, & \text{if } \dfrac{\max_{k=1}^{c} \theta_k}{\min_{k=1}^{c} \theta_k} \leq r, \\ \dfrac{\max_{k=1}^{c} \theta_k - r \min_{k=1}^{c} \theta_k}{r - 1}, & \text{otherwise,} \end{cases}$$

with a user-specified maximum weight ratio r, $r > 1$, and then Method 1 is executed with this value of the parameter b. As a consequence the cluster weights are changed only if they exceed the user-specified limit ratio.

5 Experiments

We implemented all methods suggested in the preceding section as part of an expectation maximization and fuzzy clustering program written by the first author of this paper. This program was applied to several different data sets from the UCI machine learning repository. In all data sets each dimension was normalized to mean value 0 and standard deviation 1 in order to avoid any distortions that may result from different scaling of the coordinate axes.

As one illustrative example, we present here the result of clustering the Iris data (excluding, of course, the class attribute) with the Gustafson–Kessel algorithm using three clusters of fixed size (measured as the isotropic radius) of 0.4 (since all dimensions are normalized to mean 0 and standard deviation 1, 0.4 is a good size of a cluster if three clusters are to be found). The result without shape constraints is shown in Fig. 1 at the top. Due to the few data points located in a thin diagonal cloud at the right border on the figure, the middle cluster is drawn into a fairly long and thin ellipsoid. Although this shape minimizes the objective function, it may not be a desirable result, because the cluster structure is not very compact. Using shape constraints method 2 with $r = 4$ the cluster structure shown at the bottom in Fig. 1 is obtained. In this result the clusters are more compact and resemble the class structure of the data set (to which it may be compared).

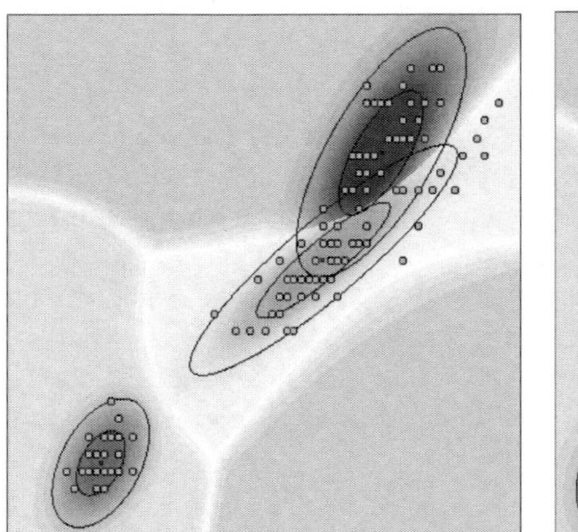
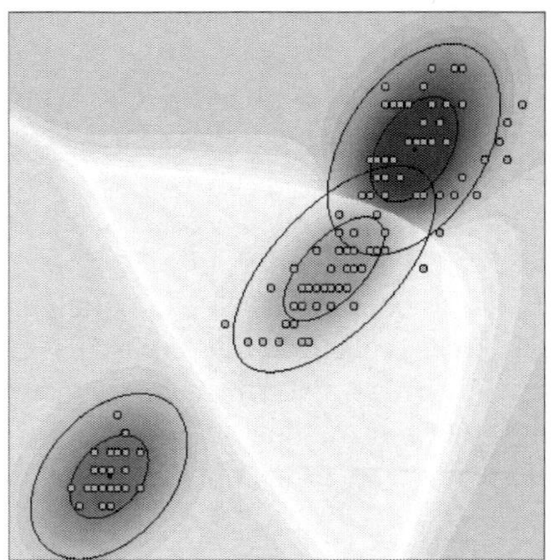

Fig. 1 Result of Gustafson–Kessel clustering on the iris data with fixed cluster size, without (*top*) and with shape regularization (*bottom*, method 2 with $r = 4$). Both images show the petal length (horizontal) and width (vertical). Clustering was done on all four attributes (sepal length and sepal width in addition to the above)

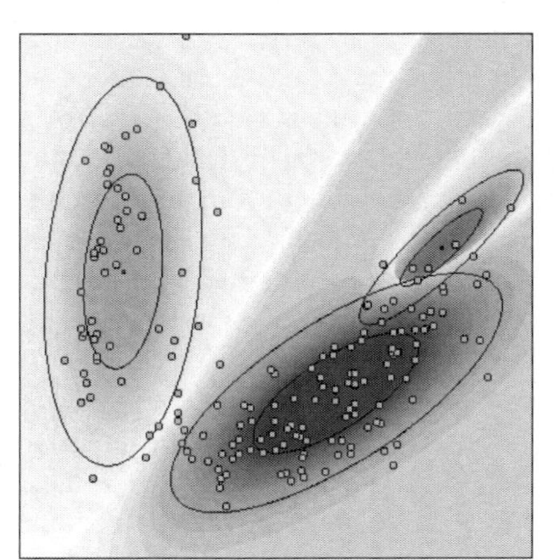
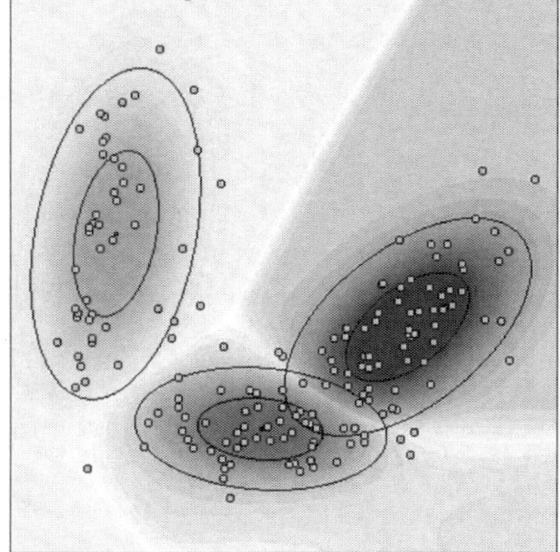

Fig. 2 Result of fuzzy maximum likelihood estimation (FMLE) algorithm on the wine data with fixed cluster weight without (*top*) and with an adaptation of (relative) cluster sizes (*bottom*, method 3 with $r = 2$). Both images show attribute 7 (horizontal) and 10 (vertical). Clustering was done on attributes 7, 10, and 13

As another example let us consider the result of clustering the wine data with the fuzzy maximum likelihood estimation (FMLE) algorithm using three clusters of variable size. We used attributes 7, 10, and 13, which are the most informative w.r.t. the class assignments. One result we obtained without constraining the relative cluster size is shown in Fig. 2 at the top. However, the algorithm is much too unstable to present a unique result. Often enough clustering fails completely, because one cluster collapses to a single data point – an effect that is mainly due to the steepness of the Gaussian probability density function and the sensitivity of the FMLE algorithm to the initialization of the cluster parameters.

This situation is considerably improved by constraining the (relative) cluster size, a result of which (that sometimes, with a fortunate initialization, can also be achieved without) is shown at the bottom in Fig. 2. It was obtained with method 3 with $r = 2$. Although the result is still not unique and sometimes clusters still focus on very few data points, the algorithm is considerably more stable and reasonable results are obtained much more often than without size constraints. Adding weight constraints (equalizing the number of data points assigned to the clusters) leads to a very stable algorithm that almost always yields the desired result. Hence we can conclude that constraining (relative) cluster size (and maybe also weight) considerably improves the robustness of the algorithm.

6 Conclusions

In this paper we suggested methods to constrain the shape as well the (relative) cluster sizc (and weight) for clustering algorithms that use cluster-specific Mahalanobis distances to describe the shape and the size of a cluster. The basic idea is to introduce a tendency towards equal length of the major axes of the represented (hyper-)ellipsoid and towards equal cluster sizes and weights. As the experiments indicate, these methods improve the robustness of the more sophisticated fuzzy clustering algorithms, which without them suffer from instabilities even on fairly simple data sets (fuzzy clustering algorithms suffer more than probabilistic methods, which seem to benefit from the "steepness" of the Gaussian function). Shape, size, and weight constrained fuzzy clustering is so robust that it can even be used without an initialization by the fuzzy c-means algorithm.

It should be noted that with a time-dependent shape constraint parameter one may obtain a soft transition from the fuzzy c-means algorithm (spherical clusters) to the Gustafson–Kessel algorithm (general ellipsoidal clusters). The fuzzy c-means algorithm results in the limit for $h \to \infty$ (or $r \to 1$). A large regularization parameter h (or an axes length ratio r close to one) enforces (almost) spherical clusters. The smaller the regularization parameter h is (or the larger the acceptable axes length ratio r), the "more ellipsoidal" or more eccentric the clusters can become. Gustafson–Kessel clustering results in the limit for $h \to 0$ (or $r \to \infty$).

Software

A free implementation of the described methods as command line programs for expectation maximization and fuzzy clustering (written in C) as well as a visualization program for clustering results can be found at

```
http://www.borgelt.net/cluster.html
http://www.borgelt.net/bcview.html
```

References

BEZDEK, J.C. (1981): *Pattern Recognition with Fuzzy Objective Function Algorithms*. Plenum, New York.

BEZDEK, J.C., KELLER, J., KRISHNAPURAM, R., and PAL, N. (1999): *Fuzzy Models and Algorithms for Pattern Recognition and Image Processing*. Kluwer, Dordrecht.

BILMES, J. (1997): A Gentle Tutorial on the EM Algorithm and Its Application to Parameter Estimation for Gaussian Mixture and Hidden Markov Models. University of Berkeley, Tech. Rep. ICSI-TR-97-021.

BLAKE, C.L. and MERZ, C.J. (1995): UCI Repository of Machine Learning Databases. http:// www.ics.uci.edu/ mlearn/MLRepository.html.

BOCK, H.H. (1974): *Automatische Klassifikation*. Vandenhoeck & Ruprecht, Göttingen, Germany.

DEMPSTER, A.P., LAIRD, N., and RUBIN, D. (1977): Maximum Likelihood from Incomplete Data via the EM Algorithm. *Journal of the Royal Statistical Society (Series B)* 39:1–38. Blackwell, Oxford, UK.

DUDA, R.O. and HART, P.E. (1973): *Pattern Classification and Scene Analysis*. Wiley, New York.

ENGL, H., HANKE, M., and NEUBAUER, A. (1996): *Regularization of Inverse Problems*. Kluwer, Dordrecht.

EVERITT, B.S. and HAND, D.J. (1981): *Finite Mixture Distributions*. Chapman & Hall, London.

GATH, I. and GEVA, A.B. (1989): Unsupervised Optimal Fuzzy Clustering. *IEEE Transactions on Pattern Analysis and Machine Intelligence* 11:773–781. IEEE, Piscataway, NJ.

GUSTAFSON, E.E. and KESSEL, W.C. (1979): Fuzzy Clustering with a Fuzzy Covariance Matrix. In *Proc. 18th IEEE Conference on Decision and Control (IEEE CDC, San Diego, CA)*, 761–766, IEEE, Piscataway, NJ.

HÖPPNER, F., KLAWONN, F., KRUSE, R., and RUNKLER, T. (1999): *Fuzzy Cluster Analysis*. Wiley, Chichester.

KELLER, A. and KLAWONN, F. (2003): Adaptation of Cluster Sizes in Objective Function Based Fuzzy Clustering. In: LEONDES, C.T. (ed.), *Database and Learning Systems IV*, 181–199. CRC, Boca Raton, FL.

KLAWONN, F. and KRUSE, R. (1997): Constructing a Fuzzy Controller from Data. *Fuzzy Sets and Systems* 85:177–193. North-Holland, Amsterdam.

KOHONEN, T. (1986): *Learning Vector Quantization for Pattern Recognition*. Technical Report TKK-F-A601. Helsinki University of Technology, Finland.

KOHONEN, T. (1995): *Self-Organizing Maps*. Springer, Heidelberg (3rd ext. edition 2001).

KRISHNAPURAM, R. and KELLER, J. (1993) A Possibilistic Approach to Clustering. *IEEE Transactions on Fuzzy Systems* 1:98–110. IEEE, Piscataway, NJ.

TIKHONOV, A.N. and ARSENIN, V.Y. (1977): *Solutions of Ill-Posed Problems*. Wiley, New York.

Dissolution and Isolation Robustness of Fixed Point Clusters

C. Hennig

Abstract The concepts of a dissolution point (which is an adaptation of the breakdown point concept to cluster analysis) and isolation robustness are introduced for general clustering methods, generating possibly overlapping clusterings. Robustness theorems for fixed point clusters (Journal of Classification 19:249–276, 2002; Journal of Multivariate Analysis 86:183–212, 2003; Data analysis and decision support, Springer, Heidelberg, 2005, pp. 47–56) are shown.

1 Introduction

Stability and robustness are important issues in cluster analysis. The addition of single outliers to the dataset can change the outcome of several cluster analysis methods completely. In Hennig (2008), a general theory for robustness and stability in cluster analysis has been introduced and discussed in depth. Several examples for lacking robustness in cluster analysis have been given, and the robustness of a wealth of cluster analysis methods including single and complete linkage, k-means, trimmed k-means and normal mixtures with fixed and estimated number of clusters has been investigated.

The present paper applies the theory given there to fixed point clusters (FPC; Hennig 2002, 2003, 2005; Hennig and Christlieb 2002).

In Sect. 2, the two robustness concepts dissolution point and isolation robustness are introduced. In Sect. 3, after fixed point clustering has been introduced, results about the dissolution point and isolation robustness of fixed point clusters are given, which are proven in Sect. 4.

C. Hennig
Department of Statistical Science, University College London, Gower St, London WC1E 6BT, UK, E-mail: chrish@stats.ucl.ac.uk

A. Okada et al. (eds.), *Cooperation in Classification and Data Analysis*, Studies in Classification, Data Analysis, and Knowledge Organizations.
DOI: 10.1007/978-3-642-00668-5_3,

2 Robustness Concepts

2.1 The Dissolution Point and a Dissimilarity Measure Between Clusters

In Hennig (2008), a definition of a breakdown point for a general clustering method has been proposed, of which the definition is based on the assignments of the points to clusters and not on parameters to be estimated. The breakdown point is a classical robustness concept, which measures the minimal amount of contamination of the dataset that suffices to drive an estimator arbitrarily far away from its initial value. There are various breakdown point definitions in the literature. The present paper makes use of the principle of the sample addition breakdown point (Donoho and Huber 1983), in which contamination is defined by adding points to the dataset. The concept introduced here deviates somewhat from the traditional meaning of the term "breakdown point", since it attributes "breakdown" to situations, which are not always the worst possible ones. Therefore, the proposed robustness measure is called "dissolution point" in the present paper, even though it is thought to measure a "breakdown" in the sense that the addition of points changes the cluster solution so strongly that the pattern of the original data can be considered as "dissolved".

A sequence of mappings $E = (E_n)_{n \in \mathbf{N}}$ is called a general clustering method (GCM), if E_n maps a set of entities $\mathbf{x}_n = \{x_1, \ldots, x_n\}$ (this is how $\mathbf{x}_n$ is always defined throughout the paper) to a collection of subsets $\{C_1, \ldots, C_k\}$ of $\mathbf{x}_n$. Note that it is assumed that entities with different indexes can be distinguished. This means that the elements of $\mathbf{x}_n$ are interpreted as data points and that $|\mathbf{x}_n| = n$ even if, for example, for $i \neq j$, $x_i = x_j$. k is not assumed to be fixed, and, as opposed to Hennig (2008), clusters are allowed to overlap.

If E is a GCM and $\mathbf{x}_{n+g}$ is generated by adding g points to $\mathbf{x}_n$, $E_{n+g}(\mathbf{x}_{n+g})$ induces a clustering on $\mathbf{x}_n$, which is denoted by $E_n^*(\mathbf{x}_{n+g})$. Its clusters are denoted by $C_1^*, \ldots, C_{k^*}^*$ (these clusters have the form $C_i \cap \mathbf{x}_n, \mathbf{C}_i \in E_{n+g}(\mathbf{x}_{n+g})$, but the notation doesn't necessarily imply $C_i \cap \mathbf{x}_n = C_i^* \forall i$). If E is a partitioning method, $E_n^*(\mathbf{x}_{n+g})$ is a partition as well. k^* may be smaller than k even if E produces k clusters for all n.

It is essential to observe that different clusters of the same clustering on the same data may have a different stability. Thus, it makes sense to define stability with respect to the individual clusters. This requires a measure for the similarity between a cluster of $E_n^*(\mathbf{x}_{n+g})$ and a cluster of $E_n(\mathbf{x}_n)$, i.e., between two subsets C and D of some finite set. For the definition of the dissolution point, the Jaccard similarity between sets is proposed:

$$\gamma(C, D) = \frac{|C \cap D|}{|C \cup D|}.$$

A similarity between C and a clustering $\hat{E}_n(\mathbf{x}_n)$ is defined by

$$\gamma^*(C, \hat{E}_n(\mathbf{x}_n)) = \min_{D \in \hat{E}_n(\mathbf{x}_n)} \gamma(C, D).$$

Definition 1. Let $E = (E_n)_{n \in \mathbb{N}}$ be a GCM. The *dissolution point* of a cluster $C \in E_n(\mathbf{x}_n)$ is defined as

$$\Delta(E, \mathbf{x}_n, C) = \min_g \left\{ \frac{g}{|C| + g} : \exists \mathbf{x}_{n+g} = (x_1, \ldots, x_{n+g}) : \right.$$
$$\left. \gamma^*(C, E_n^*(\mathbf{x}_{n+g})) \leq \frac{1}{2} \right\}.$$

The choice of the cutoff value $\frac{1}{2}$ is motivated in Hennig (2006).

2.2 Isolation Robustness

Results on dissolution points are informative about the methods, but they do not necessarily allow a direct comparison of different methods, because they usually need method-specific assumptions. The concept of isolation robustness is thought to enable such a comparison. The rough idea is that it can be seen as a minimum robustness requirement of cluster analysis that an extremely well isolated cluster remains stable under the addition of points. The isolation $i(C)$ of a cluster C is defined as the minimum distance of a point of the cluster to a point not belonging to the cluster, which means that a distance structure on the data is needed.

The definition of isolation robustness in Hennig (2008) applies to partitioning methods only. Fixed point clustering is not a partitioning method, and therefore a new definition, which I call "of type II", is presented here. Let $\mathcal{M}_m$ be the space of distance matrices between m objects permissible by the distance structure underlying the GCM. Call a cluster $C \in E_n(\mathbf{x})$ "fulfilling the isolation condition for a function v_m" (v_m is assumed to map $\mathcal{M}_m \times \mathbb{N} \mapsto I\!R$) if $|C| = m$, M_C is its within-cluster distance matrix, $i(C) > v_m(M_C, g)$, and an additional homogeneity condition is fulfilled, which may depend on the GCM.

Definition 2. A GCM $E = (E_n)_{n \in \mathbb{N}}$ is called *isolation robust* of type II, if there exists a sequence of functions v_m, $m \in \mathbb{N}$ such that for $n \geq m$ for any dataset $\mathbf{x}_n$, for given $g \in \mathbb{N}$, for any cluster $C \in E_n(\mathbf{x})$ fulfilling the isolation condition for v_m, and for any dataset $\mathbf{x}_{n+g}$, where g points are added to $\mathbf{x}_n$, there exists $D \in E_n^*(\mathbf{x}_{n+g}) : D \subseteq C, \gamma(C, D) > \frac{1}{2}$.

The reason that the isolation condition requires a further unspecified homogeneity condition is that extreme isolation of clusters should prevent by no means that a cluster is split up into several smaller clusters by the addition of points. This is not considered by the isolation robustness concept, and therefore a further condition on C has to ensure that this does not happen.

3 Fixed Point Clusters

3.1 Definition of Fixed Point Clusters

FPC analysis has been introduced as a method for overlapping clustering for clusterwise linear regression (Hennig 2002, 2003) and normal-shaped clusters of p-dimensional data (Hennig 2005; Hennig and Christlieb 2002), which should be robust against outliers.

The basic idea of FPC analysis is that a cluster can be formalized as a data subset, which is homogeneous in the sense that it does not contain any outlier, and which is well separated from the rest of the data meaning that all other points are outliers with respect to the cluster. That is, the FPC concept is a local cluster concept: It does not assume a cluster structure or some parametric model for the whole dataset. It is based only on a definition of outliers with respect to the cluster candidate itself.

In order to define FPCs, a definition of an outlier with respect to a data subset is needed. The definition should be based only on a parametric model for the non-outliers (reference model), but not for the outliers. That is, if the Gaussian family is taken as reference model, the whole dataset is treated as if it came from a contamination mixture

$$(1-\epsilon)N_p(a,\mathbf{\Sigma})+\epsilon P^*, \quad 0\leq\epsilon<1, \tag{1}$$

where p is the number of variables, $N_p(a,\mathbf{\Sigma})$ denotes the p-dimensional Gaussian distribution with mean vector a and covariance matrix $\mathbf{\Sigma}$, and P^* is assumed to generate points well separated from the core area of $N_p(a,\mathbf{\Sigma})$. The principle to define the outliers is taken from Becker and Gather (1999). They define α-outliers as points that lie in a region with low density such that the probability of the so-called outlier region is α under the reference distribution. α has to be small in order to match the impression of outlyingness. For the $N_p(a,\mathbf{\Sigma})$-distribution, the α-outlier region is

$$\{x:\ (x-a)^t\mathbf{\Sigma}^{-1}(x-a)>\chi^2_{p;1-\alpha}\},$$

$\chi^2_{p;1-\alpha}$ denoting the $1-\alpha$-quantile of the χ^2-distribution with p degrees of freedom.

Note that it is not assumed that the whole dataset can be partitioned into clusters of this kind, and therefore this does not necessarily introduce a parametric model for the whole dataset.

In a concrete situation, a and $\mathbf{\Sigma}$ are not known, and they have to be estimated. This is done for Mahalanobis FPCs by the sample mean and the maximum likelihood covariance matrix. (Note that these estimators are non-robust, but they are reasonable if they are only applied to the non-outliers.)

A dataset $\mathbf{x}_n$ consists of p-dimensional points. Data subsets are represented by an indicator vector $w\in\{0,1\}^n$. Let $\mathbf{x}_n(w)$ be the set with only the points x_i, for which $w_i=1$, and $n(w)=\sum_{i=1}^n w_i$. Let $m(w)=\frac{1}{n(w)}\sum_{w_i=1}x_i$ the mean vector and

$\mathbf{S}(w) = \frac{1}{n(w)} \sum_{w_i=1} (x_i - m(w))(x_i - m(w))'$ the ML covariance matrix estimator for the points indicated by w.

The set of outliers from $\mathbf{x}_n$ with respect to a data subset $\mathbf{x}_n(w)$ is

$$\{x : (x - m(w))'\mathbf{S}(w)^{-1}(x - m(w)) > \chi^2_{p;1-\alpha}\}.$$

That is, a point is defined as an outlier w.r.t $\mathbf{x}_n(w)$, if its Mahalanobis distance to the estimated parameters of $\mathbf{x}_n(w)$ is large.

An FPC is defined as a data subset which is exactly the set of non-outliers w.r.t. itself:

Definition 3. A data subset $\mathbf{x}_n(w)$ of $\mathbf{x}_n$ is called *Mahalanobis fixed point cluster* of level α, if for $i = 1, \dots, n$:

$$w = \left(1\left[(x_i - m(w))'\mathbf{S}(w)^{-1}(x_i - m(w)) \leq \chi^2_{p;1-\alpha}\right]\right)_{i=1,\dots,n}. \quad (2)$$

If $\mathbf{S}(w)^{-1}$ does not exist, the Moore–Penrose inverse is taken instead on the supporting hyperplane of the corresponding degenerated normal distribution, and $w_i = 0$ for all other points (the degrees of freedom of the χ^2-distribution may be adapted in this case).

For combinatorial reasons it is impossible to check (2) for all w. But FPCs can be found by a fixed point algorithm defined by

$$w^{k+1} = \left(1\left[(x_i - m(w^k))'\mathbf{S}(w^k)^{-1}(x_i - m(w^k)) \leq \chi^2_{p;1-\alpha}\right]\right)_{i=1,\dots,n}. \quad (3)$$

This algorithm is shown to converge toward an FPC in a finite number of steps if $\chi^2_{p;1-\alpha} > p$ (which is always fulfilled for $\alpha < 0.25$, i.e., for all reasonable choices of α) in Hennig and Christlieb (2002).

The problem here is the choice of reasonable starting configurations w^0. While, according to this definition, there are many very small FPCs, which are not very meaningful (e.g., all sets of p or fewer points are FPCs), an FPC analysis aims at finding all substantial FPCs, where "substantial" means all FPCs corresponding to well separated, not too small data subsets which give rise to an adequate description of the data by a model of the form (1). For clusterwise regression, this problem is discussed in depth in Hennig (2002) along with an implementation, which is included in the add-on package "fpc" for the statistical software system R. In the same package, there is also an implementation of Mahalanobis FPCs. There, the following method to generate initial subsets is applied.

For every point of the dataset, one initial configuration is chosen, so that there are n runs of the algorithm (3). For every point, the p nearest points in terms of the Mahalanobis distance w.r.t. $\mathbf{S}(1, \dots, 1)$ are added, so that there are $p + 1$ points. Because such configurations often lead to too small clusters, the initial configuration is enlarged to contain n_{start} points. To obtain the $p + 2$nd to the n_{start}th point, the covariance matrix of the current configuration is computed (new for every added point) and the nearest point in terms of the new Mahalanobis distance is added.

$n_{\text{start}} = 20 + 4p$ is chosen as the default size of initial configurations in package "fpc". This is reasonable for fairly large datasets, but should be smaller for small datasets. Experience shows that the effective minimum size of FPCs that can be found by this method is not much smaller than n_{start}. The default choice for α is 0.99; $\alpha = 0.95$ produces in most cases more FPCs, but these are often too small, compare Example 1. Note that Mahalanobis FPCs are invariant under linear transformations.

3.2 Robustness Results for Fixed Point Clusters

To derive a lower bound for the dissolution point of a fixed point cluster, the case $p = 1$ is considered for the sake of simplicity. This is a special case of both Mahalanobis and clusterwise linear regression FPCs (intercept only).

$E_n(\mathbf{x}_n)$ is taken as the collection of all data subsets fulfilling (2). $E_n(\mathbf{x}_n)$ is not necessarily a partition, because FPCs may overlap and not all points necessarily belong to any FPC.

FPCs are robust against gross outliers in the sense that
an FPC $\mathbf{x}(w)$ is invariant against any change, especially addition of points,

$$\text{outside its domain } \{(x - m(w))'\mathbf{S}(w)^{-1}(x - m(w)) \leq c\},\ c = \chi^2_{p;1-\alpha}, \tag{4}$$

because such changes simply do not affect its definition. However, FPCs can be affected by points added inside their domain, which is, for $p = 1$,

$$D(w) = [m(w) - s(w)\sqrt{c}, m(w) + s(w)\sqrt{c}],\ s(w) = \sqrt{S(w)}.$$

The aim of the following theory is to characterize a situation in which an FPC is stable under addition of points. The key condition is the separateness of the FPC, i.e., the number of points in its surrounding [which is bounded by k_2 in (7)] and the number of points belonging to it but close to its border [which is bounded by k_1 in (6)]. The derived conditions for robustness (in the sense of a lower bound on the dissolution point) are somewhat stronger than presumably needed, but the theory reflects that the key ingredient for stability of an FPC is to have few points close to the border (inside and outside).

In the following, $\mathbf{x}_n(w)$ denotes a Mahalanobis FPC in $\mathbf{x}_n$.

Let $S_{gk}(w)$ be the set containing the vectors $(m_{+g}, s^2_{+g}, m_{-k}, s^2_{-k})$ with the following property:

Property $A(g, k, \mathbf{x}_n(w))$: Interpret $\mathbf{x}_n(w)$ as an FPC on itself as complete dataset, i.e., on $\mathbf{y}_{\tilde{n}} = \mathbf{x}_n(w) = \mathbf{y}_{\tilde{n}}(1, \ldots, 1)$ $(\tilde{n} = n(w))$.
$(m_{+g}, s^2_{+g}, m_{-k}, s^2_{-k})$ has the Property $A(g, k, \mathbf{x}_n(w))$ if there exist points $y_{\tilde{n}+1}, \ldots, y_{\tilde{n}+g}$ such that, if the algorithm (3) is run on the dataset $\mathbf{y}_{\tilde{n}+g} = \mathbf{y}_{\tilde{n}} \cup \{y_{\tilde{n}+1}, \ldots, y_{\tilde{n}+g}\}$ and started from the initial dataset $\mathbf{y}_{\tilde{n}}$, it converges to a new FPC $\mathbf{y}_{\tilde{n}+g}(w^*)$ such that m_{+g} and s^2_{+g} are the values of the mean and variance

of the points $\{y_{\tilde{n}+1}, \dots, y_{\tilde{n}+g}\} \cap \mathbf{y}_{\tilde{n}+g}(w^*)$, and m_{-k} and s^2_{-k} are the values of the mean and variance of the points lost in the algorithm, i.e., $\mathbf{y}_{\tilde{n}} \setminus \mathbf{y}_{\tilde{n}+g}(w^*)$ (implying $|\mathbf{y}_{\tilde{n}} \setminus \mathbf{y}_{\tilde{n}+g}(w^*)| \leq k$). Mean and variance of 0 points are taken to be 0. Note that always $(0,0,0,0) \in S_{gk}(w)$, because of (4) and the added points can be chosen outside the domain of $\mathbf{y}_{\tilde{n}}$.

In the proof of Theorem 1 it will be shown that an upper bound of the domain of $\mathbf{y}_{\tilde{n}+g}(w^*)$ in the situation of Property $A(g,k,\mathbf{x}_n(w))$ (assuming $m(w) = 0,\ s(w) = 1$) is

$$\mathbf{x}_{\max}(g,k,m_{+g},s^2_{+g},m_{-k},s^2_{-k})$$
$$= \frac{n_g m_{+g} - k m_{-k}}{n_1} + \sqrt{c\left(\frac{n(w) + n_g s^2_{+g} - k s^2_{-k}}{n_1} + \frac{c_1 m^2_{+g} + c_2 m^2_{-k} + c_3 m_{+g} m_{-k}}{n_1^2}\right)}, \quad (5)$$

where $n_g = |\{w^*_j = 1 : \ j \in \{n+1, \dots, n+g\}\}|$ is the number of points added during the algorithm,

$$n_1 = n(w) + n_g - k, \quad c_1 = (n(w) - k) n_g, \quad c_2 = -(n(w) + n_g)k, \quad c_3 = 2k n_g.$$

Further define for $g, k \geq 0$

$$x_{\mathrm{maxmax}}(g,k) = \max_{(m_{+g}, s^2_{+g}, m_{-k}, s^2_{-k}) \in S_{gk}(w)} \mathbf{x}_{\max}(g,k,m_{+g},s^2_{+g},m_{-k},s^2_{-k}),$$
$$x_{\mathrm{maxmin}}(g,k) = \min_{(m_{+g}, s^2_{+g}, m_{-k}, s^2_{-k}) \in S_{gk}(w)} \mathbf{x}_{\max}(g,k,m_{+g},s^2_{+g},m_{-k},s^2_{-k}).$$

Note that $x_{\mathrm{maxmin}}(g,k) \leq \sqrt{c} \leq x_{\mathrm{maxmax}}(g,k)$, because $(0,0,0,0) \in S_{gk}(w)$. $x_{\mathrm{maxmax}}(g,k)$ is nondecreasing in g, because points can always be added far away that they do not affect the FPC, and therefore a maximum for smaller g can always be attained for larger g. By analogy, $x_{\mathrm{maxmin}}(g,k)$ is non-increasing.

Theorem 1. *Let $\mathbf{x}_n(w)$ be an FPC in $\mathbf{x}_n$. Let $\mathbf{x}_{n+g} = \{x_1, \dots, x_{n+g}\}$. If $\exists k_1, k_2$ with*

$$\begin{aligned} k_1 \leq |\mathbf{x}_n \cap & \\ & ([m(w) - s(w)x_{\mathrm{maxmax}}(g+k_1,k_2), m(w) - s(w)\sqrt{c}] \cup \\ & [m(w) + s(w)\sqrt{c}, m(w) + s(w)x_{\mathrm{maxmax}}(g+k_1,k_2)])|, \end{aligned} \quad (6)$$
$$\begin{aligned} k_2 \leq |\mathbf{x}_n \cap & \\ & ([m(w) - s(w)\sqrt{c}, m(w) - s(w)x_{\mathrm{maxmin}}(g+k_1,k_2)] \cup \\ & [m(w) + s(w)x_{\mathrm{maxmin}}(g+k_1,k_2), m(w) + s(w)\sqrt{c}])|, \end{aligned} \quad (7)$$

then

$$\gamma^*(\mathbf{x}(w), E^*_n(\mathbf{x}_{n+g})) \geq \frac{n(w) - k_2}{n(w) + k_1}. \quad (8)$$

If $\frac{n(w)-k_2}{n(w)+k_1} > \frac{1}{2}$, then $\Delta(\mathbf{x}_n(w), \mathbf{x}_n) > \frac{g}{n(w)+g}$.

The proof is given in Sect. 4. k_1 is the maximum number of points in $\mathbf{x}_n$ outside the FPC $\mathbf{x}_n(w)$ that can be added during the algorithm, k_2 is the maximum number of points inside the FPC $\mathbf{x}_n(w)$ that can be lost during the algorithm due to changes caused by the g new points.

Theorem 1 shows the structure of the conditions needed for stability, but in the given form it is not obvious how strong these conditions are (and even not if they are possible to fulfill) for a concrete dataset. It is difficult to evaluate $x_{\text{maxmax}}(g+k_1, k_2)$ and $x_{\text{maxmin}}(g+k_1, k_2)$ and the conditions (6) and (7), where k_1 and k_2 also appear on the right hand sides. The following Lemma will give somewhat conservative bounds for $x_{\text{maxmax}}(g+k_1, k_2)$ and $x_{\text{maxmin}}(g+k_1, k_2)$ which can be evaluated more easily and will be applied in Example 1. The conditions (6) and (7) can then be checked for any given g, k_1 and k_2.

Lemma 1. *For* $g \geq 0, 0 \leq k < n(w)$:

$$x_{\text{maxmax}}(g,k) \leq x^*_{\text{maxmax}}(g,k,m^*_{+g}), \tag{9}$$

$$x_{\text{maxmin}}(g,k) \geq x^*_{\text{maxmin}}(g,k,m^*_{-k}), \tag{10}$$

where for $0 \leq k < n(w)$

$$x^*_{\text{maxmax}}(0,k,m_{+g}) = \sqrt{c}, \textit{ for } g > 0:$$

$$x^*_{\text{maxmax}}(g,k,m_{+g}) = \frac{g m_{+g} + k\sqrt{c}}{n_1} + \sqrt{c\left(\frac{n(w) + g(a_{\max}(g)^2 - m^2_{+g})}{n_1} + \frac{c_1 m^2_{+g}}{n_1^2}\right)},$$

$$x^*_{\text{maxmin}}(g,k,m_{-k}) = \frac{-g m^*_{+g} - k m_{-k}}{n_1} + \sqrt{c\left(\frac{n(w) - k(c - m^2_{-k})}{n_1} + \frac{c_2 m^2_{-k} - c_3 m_{-k} m^*_{+g}}{n_1^2}\right)},$$

$$a_{\max}(g) = x^*_{\text{maxmax}}(g-1,k,m^*_{+(g-1)}),$$

$$m^*_{+g} = \frac{1}{g}\sum_{i=1}^{g} a_{\max}(i),$$

$$m^*_{-k} = \underset{m_{-k}\in[0,\sqrt{c}]}{\arg\min}\; x^*_{\text{maxmin}}(g,k,m_{-k}),$$

The proof is given in Sect. 4. For the minimization needed to obtain m^*_{-k}, the zeros of the derivative of $x^*_{\text{maxmin}}(g,k,m_{-k})$ are the zeros of $t m^2_{-k} + u m_{-k} + v$ where

$$t = k^3 + \frac{c^2}{n_1^2} - n_1 ck^2 + 2k^2 c(n(w) + g) - \frac{k^2(n(w) + g)^2 c}{n_1},$$
$$u = -\frac{2kgm^*_{+g}}{n_1} + 2k^2 gcm^*_{+g} - \frac{2k^2(n(w) + g)gcm^*_{+g}}{n_1},$$
$$v = k^2 n(w) - k^3 c + \frac{(n(w) - k)g(m^*_{+g})^2}{n_1} - \frac{k^2 g^2 c(m^*_{+g})^2}{n_1}. \quad (11)$$

Theorem 2. *FPC analysis is isolation robust of type II under the following condition on an FPC* $C = \mathbf{x}_n(w)$ *with* $i(C) > v_m(M_C, g)$:

$$\exists k_2 : \frac{|C| - k_2}{|C|} > \frac{1}{2},$$
$$k_2 \leq |T(C) \cap ([-s(w)\sqrt{c}, -s(w)x_{\text{maxmin}}(g, k_2)] \cup [s(w)x_{\text{maxmin}}(g, k_2), s(w)\sqrt{c}])|, \quad (12)$$

where $T(C) = \mathbf{x}_n(w) - m(w)$ *is* C *transformed to mean 0.*

Example 1. The dataset shown in Fig. 1 consists of two datasets of the form $\Phi^{-1}_{a,\sigma^2}(\frac{1}{n+1}), \ldots, \Phi^{-1}_{a,\sigma^2}(\frac{n}{n+1})$ with $n = 25$, $(a, \sigma^2) = (0, 1)$, $(5, 1)$, respectively. For $\alpha = 0.99$, the computation scheme outlined in Sect. 3.1 finds two FPCs, namely the two separated initial datasets, for n_{start} down to 4. Let $\mathbf{x}_n(w)$ be the 25 points generated with $(a, \sigma^2) = (0, 1)$, $m(w) = 0$, $s(w)^2 = 0.788$, $D(w) = [-2.287, 2.287]$. The largest point is 1.769, the second largest one is 1.426, the smallest point in the data not belonging to $\mathbf{x}(w)$ is 3.231, the second smallest one is 3.574. If $g = 1$ point is added, $s(w)x^*_{\text{maxmax}}(1, 0, m^*_{+1}) = 2.600$, $s(w)x^*_{\text{maxmin}}(1, 0, m^*_{-0}) = 2.154$. Thus, (6) and (7) hold for $k_1 = k_2 = 0$. The same holds for $g = 2$: $s(w)x^*_{\text{maxmax}}(2, 0, m^*_{+2}) = 3.000$, $s(w)x^*_{\text{maxmin}}(2, 0, m^*_{-0}) = 2.019$. For $g = 3$: $s(w)x^*_{\text{maxmax}}(3, 0, m^*_{+3}) = 3.528$, $s(w)x^*_{\text{maxmin}}(3, 0, m^*_{-0}) = 1.879$. This means that (6) does not hold for $k_1 = 0$, because the smallest point belonging to $(a, \sigma^2) = (5, 1)$ would be included into the corresponding FPC. $g = 3$ and $k_1 = 1$ in Theorem 1 correspond to $g = 4$ in Lemma 1. For $g = 4$: $s(w)x^*_{\text{maxmax}}(4, 0, m^*_{+4}) = 4.250$, $s(w)x^*_{\text{maxmin}}(4, 0, m^*_{-0}) = 1.729$. This means that for $g = 3$, neither $k_1 = 1$, nor $k_2 = 0$ works, and in fact an iteration of (3) with added points $2.286, 2.597, 2.929$ leads to dissolution, namely to an FPC containing all 50 points of the dataset. Thus, $\Delta(E_n, \mathbf{x}_n, \mathbf{x}_n(w)) = \frac{3}{28}$.

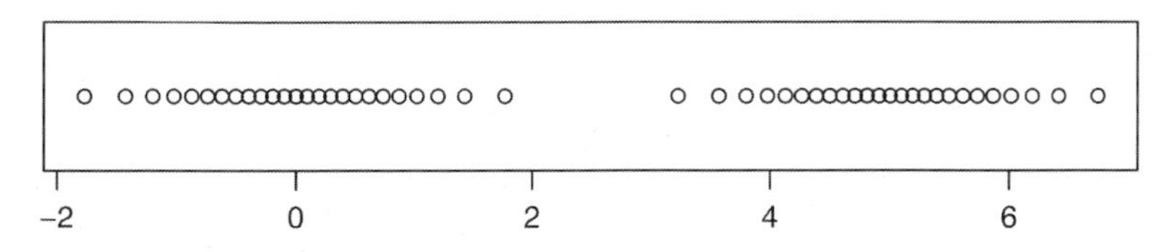

Fig. 1 Example dataset

For $\alpha = 0.95$, there is also an FPC $\mathbf{x}_n(w_{0.95})$ corresponding to $(a, \sigma^2) = (0, 1)$, but it only includes 23 points, the two most extreme points on the left and on the right are left out. According to the theory, this FPC is not dissolved by being joined with the points corresponding to $(a, \sigma^2) = (5, 1)$, but by implosion. For $g = 1$, $s(w_{0.95})x^*_{\text{maxmax}}(1, 0, m^*_{+1}) = 1.643$, $s(w_{0.95})x^*_{\text{maxmin}}(1, 0, m^*_{-0}) = 1.405$. This means that the points $-1.426, 1.426$ can be lost. $s(w_{0.95})x^*_{\text{maxmax}}(1, 2, m^*_{+1}) = 1.855$, $s(w_{0.95})x^*_{\text{maxmin}}(1, 2, m^*_{-2}) = 0.988$, which indicates that k_2 is still too small for (7) to hold. Nothing better can be shown than $\Delta(E_{n,0.05}, \mathbf{x}_n, \mathbf{x}_n(w_{0.05})) \geq \frac{1}{24}$. However, here the conservativity of the dissolution bound matters (the worst case of the mean and the variance of the two left out points used in the computation of $x^*_{\text{maxmin}}(1, 2, m^*_{-2})$ cannot be reached at the same time in this example) and dissolution by addition of one (or even two) points seems to be impossible.

4 Proofs

Proof of Theorem 1. Because of the invariance of FPCs and the equivariance of their domain under linear transformations, assume w.l.o.g. $m(w) = 0,\ s(w) = 1$.

First it is shown that $\mathbf{x}_{\max}(g, k, m_{+g}, s^2_{+g}, m_{-k}, s^2_{-k})$ as defined in (5) is the upper bound of the domain of $\mathbf{y}_{\tilde{n}+g}(w^*)$ in the situation of Property $A(g, k, \mathbf{x}_n(w))$, i.e.,

$$\mathbf{x}_{\max}(g, k, m_{+g}, s^2_{+g}, m_{-k}, s^2_{-k}) = m(w^*) + \sqrt{c}s(w^*). \tag{13}$$

Assume, w.l.o.g., that the k points to be lost during the algorithm are $y_1, \ldots, y_k$ and the n_g added points are $y_{\tilde{n}+1}, \ldots, y_{\tilde{n}+n_g}$, thus $\mathbf{y}_{\tilde{n}+g}(w^*) = \{y_{k+1}, \ldots, y_{\tilde{n}+n_g}\}$, $|\mathbf{y}_{\tilde{n}+g}(w^*)| = n_1$. Now, by straightforward arithmetic:

$$m(w^*) = \frac{n(w)m(w) + n_g m_{+g} - k m_{-k}}{n_1} = \frac{n_g m_{+g} - k m_{-k}}{n_1},$$

$$\begin{aligned} s(w^*)^2 &= \frac{1}{n_1}\left(\sum_{i=1}^{\tilde{n}}(y_i - m(w^*))^2 + \sum_{w_i^*=1, w_i=0}(y_i - m(w^*))^2 \right.\\ &\quad \left. - \sum_{i=1}^{k}(y_i - m(w^*))^2\right) \\ &= \frac{1}{n_1}\left[\sum_{i=1}^{\tilde{n}}\left(y_i - \frac{n_g m_{+g} - k m_{-k}}{n_1}\right)^2 \right.\\ &\quad + \sum_{w_i^*=1, w_i=0}\left(y_i - m_{+g} + \frac{(n(w) - k)m_{+g} + k m_{-k}}{n_1}\right)^2 \\ &\quad \left. - \sum_{i=1}^{k}\left(y_i - m_{-k} + \frac{(n(w) + n_g)m_{-k} - n_g m_{+g}}{n_1}\right)^2\right] \end{aligned}$$

$$= \frac{n(w)s(w)^2 + n_g s_{+g}^2 - ks_{-k}^2}{n_1}$$
$$+ \frac{1}{n_1^3}[(n(w)n_g^2 + n_g(n(w)-k)^2 - kn_g^2)m_{+g}^2$$
$$+ (n(w)k^2 + n_g k^2 - (n(w)+n_g)^2 k)m_{-k}^2$$
$$+ (2kn_g(n(w)-k) - 2n(w)kn_g + 2kn_g(n(w)+n_g))m_{+g}m_{-k}].$$

This proves (13).

It remains to prove (8), then the bound on Δ follows directly from Definition 1. From (13), get that in the situation of Property $A(g,k,\mathbf{x}_n(w))$, the algorithm (3), which is known to converge, will always generate FPCs in the new dataset $\mathbf{y}_{\tilde{n}+g}$ with a domain $[x^-, x^+]$, where

$$x^- \in [-x_{\text{maxmax}}(g,k), -x_{\text{maxmin}}(g,k)],\ x^+ \in [x_{\text{maxmin}}(g,k), x_{\text{maxmax}}(g,k)], \quad (14)$$

if started from $\mathbf{x}_n(w)$. Note that, because of (4), the situation that $\mathbf{x}_n(w) \subset \mathbf{x}_n$ is FPC, g points are added to $\mathbf{x}_n$, k_1 further points of $\mathbf{x}_n \setminus \mathbf{x}_n(w)$ are included in the FPC and k_2 points from $\mathbf{x}_n(w)$ are excluded during the algorithm (3) is equivalent to the situation of property $A(g+k_1,k_2)$. Compared to $\mathbf{x}(w)$, if g points are added to the dataset and no more than k_1 points lie in $[-x_{\text{maxmax}}(g+k_1,k_2), -\sqrt{c}] \cup [\sqrt{c}, x_{\text{maxmax}}(g+k_1,k_2)]$, no more than $g+k_1$ points can be added to the original FPC $\mathbf{x}_n(w)$. Only the points of $\mathbf{x}_n(w)$ in $[-\sqrt{c}, -x_{\text{maxmin}}(g+k_1,k_2)] \cup [x_{\text{maxmin}}(g+k_1,k_2), \sqrt{c}]$ can be lost. Under (7), these are no more than k_2 points, and under (6), no more than k_1 points of $\mathbf{x}_n$ can be added. The resulting FPC $\mathbf{x}_{n+g}(w^*)$ has in common with the original one at least $n(w) - k_2$ points, and $|\mathbf{x}_n(w) \cup (\mathbf{x}_n \cap \mathbf{x}_{n+g}(w^*))| \leq n(w) + k_1$, which proves (8).

The following proposition is needed to show Lemma 1:

Proposition 1. *Assume* $k < n(w)$. *Let* $\mathbf{y} = \{x_{n+1}, \ldots, x_{n+g}\}$. *In the situation of Property* $A(g,k,\mathbf{x}_n(w))$, $m_{+g} \leq m_{+g}^*$ *and* $\max(\mathbf{y} \cap \mathbf{x}_{n+g}(w^*)) \leq a_{\max}(g)$.

Proof by induction over g.
$g = 1$: $x_{n+1} \leq \sqrt{c}$ is necessary because otherwise the original FPC would not change under (3).
$g > 1$: suppose that the proposition holds for all $h < g$, but not for g. There are two potential violations of the proposition, namely $m_{+g} > m_{+g}^*$ and $\max(\mathbf{y} \cap \mathbf{x}_{n+g}(w^*)) > a_{\max}(g)$. The latter is not possible, because in previous iterations of (3), only $h < g$ points of $\mathbf{y}$ could have been included, and because the proposition holds for h, no point larger than $x_{\text{maxmax}}^*(g-1,k,m_{+(g-1)}^*)$ can be reached by (3). Thus, $m_{+g} > m_{+g}^*$. Let w.l.o.g. $x_{n+1} \leq \ldots \leq x_{n+g}$. There must be $h < g$ so that $x_{n+h} > a_{\max}(h)$. But then the same argument as above excludes that x_{n+h} can be reached by (3). Thus, $m_{+g} \leq m_{+g}^*$, which proves the proposition.

Proof of Lemma 1. Proof of (9): Observe that $\mathbf{x}_{\max}(g,k,m_{+g},s_{+g}^2,m_{-k},s_{-k}^2)$ is enlarged by setting $s_{-k}^2 = 0, \mathbf{n}_g = g$ and by maximizing s_{+g}^2. $s_{+g}^2 \leq a_{\max}(g)^2 - m_{+g}^2$ because of Proposition 1. Because $x_{\text{maxmax}}(g,k) \geq \sqrt{c}$, if

$\mathbf{x}_{\max}(g, k, m_{+g}, s_{+g}^2, m_{-k}, s_{-k}^2)$ is maximized in a concrete situation, the points to be left out of $\mathbf{x}_n(w)$ must be the smallest points of $\mathbf{x}_n(w)$. Thus, $-\sqrt{c} \leq m_{-k} \leq m(w) = 0$.

Further, $c_2 \leq 0, c_3 \geq 0$. To enlarge $\mathbf{x}_{\max}(g, k, m_{+g}, s_{+g}^2, m_{-k}, s_{-k}^2)$, replace the term $-km_{-k}$ in (5) by $k\sqrt{c}$, $c_2 m_{-k}^2$ by 0 and $c_3 m_{+g} m_{-k}$ by 0 (if $m_{+g} < 0$ then $m_{-k} = 0$, because in that case m_{+g} would enlarge the domain of the FPC in both directions and $\mathbf{x}_{n+g}(w^*) \supseteq \mathbf{x}_n(w)$). By this, obtain $x^*_{\text{maxmax}}(g, k, m_{+g})$, which is maximized by the maximum possible m_{+g}, namely m^*_{+g} according to Proposition 1.

Proof of (10). To reduce $\mathbf{x}_{\max}(g, k, m_{+g}, s_{+g}^2, m_{-k}, s_{-k}^2)$, set $s_{+g}^2 = 0$ and observe $s_{-k}^2 \leq (c - m_{-k}^2)$. The minimizing m_{-k} can be assumed to be positive (if it would be negative, $-m_{-k}$ would yield an even smaller $\mathbf{x}_{\max}(g, k, m_{+g}, s_{+g}^2, m_{-k}, s_{-k}^2))$. $c_1 \geq 0$, and therefore $n_g m_{+g}$ can be replaced by $-gm^*_{+g}$, $c_1 m_{+g}^2$ can be replaced by 0, and $c_3 m_{-k} m_{+g}$ can be replaced by $-c_3 m_{-k} m^*_{+g}$. This yields (10).

Proof of Theorem 2. Note first that for one-dimensional data, $T(C)$ can be reconstructed from the distance matrix M_C. If $i(C) > s(w)\mathbf{x}_{\text{maxmax}}(g, k_2)$, there are no points in the transformed dataset $\mathbf{x}_n - m(w)$ that lie in

$$([-s(w)\mathbf{x}_{\text{maxmax}}(g, k_2), -s(w)\sqrt{c}] \cup [s(w)\sqrt{c}, s(w)\mathbf{x}_{\text{maxmax}}(g, k_2)]),$$

and it follows from Theorem 1 that

$$\exists D \in E_n^*(\mathbf{x}_{n+g}) : \ D \subseteq C, \ \gamma(C, D) \geq \frac{|C| - k_2}{|C|} > \frac{1}{2}.$$

$v_m(M_C, g) = s(w)\mathbf{x}_{\text{maxmax}}(g, k_2)$ is finite by Lemma 1 and depends on C and $\mathbf{x}_n$ only through $s(w)$ and k_2, which can be determined from M_C.

References

BECKER, C. and GATHER, U. (1999): The Masking Breakdown Point of Multivariate Outlier Identification Rules. *Journal of the American Statistical Association, 94,* 947–955.

DONOHO, D. L. and HUBER, P. J. (1983): The Notion of Breakdown Point. In: P. J. Bickel, K. Doksum, and J. L. Hodges jr., J. L. (Eds.): *A Festschrift for Erich L. Lehmann*, Wadsworth, Belmont, CA, 157–184.

HENNIG, C. (2002): Fixed Point Clusters for Linear Regression: Computation and Comparison. *Journal of Classification 19,* 249–276.

HENNIG, C. (2003): Clusters, Outliers, and Regression: Fixed Point Clusters. *Journal of Multivariate Analysis, 86,* 183–212.

HENNIG, C. (2005): Fuzzy and Crisp Mahalanobis Fixed Point Clusters. In: D. Baier, R. Decker, and L. Schmidt-Thieme (Eds.): *Data Analysis and Decision Support*. Springer, Heidelberg, 47–56.

HENNIG, C. (2008): Dissolution Point and Isolation Robustness: Robustness Criteria for General Cluster Analysis Methods. *Journal of Multivariate Analysis, 99(6),* 1154–1176. DOI: 10.1016/j.jmva.2007.07.002.

HENNIG, C. and CHRISTLIEB, N. (2002): Validating visual clusters in large datasets: fixed point clusters of spectral features. *Computational Statistics and Data Analysis, 40,* 723–739.

ADCLUS: A Data Model for the Comparison of Two-Mode Clustering Methods by Monte Carlo Simulation

M. Wiedenbeck and S. Krolak-Schwerdt

Abstract In this paper, it is demonstrated that the generalized ADCLUS model (Psychological Review 86:87–123, 1979; Psychometrika 47:449–475, 1982) may serve as a theoretical data model for two-mode clustering. The data model has the potential to offer a generale rationale for the generation of artificial data sets in Monte Carlo experiments in that a number of model parameters are included which may generate clusters of different shape, overlap, between-group heterogeneity, etc. The usefulness of the data model as a framework to comparatively estimate the performance of some two-mode methods is demonstrated in a Monte Carlo study.

1 Introduction

This paper is concerned with the cluster analysis of two-mode data which consist of two sets of entities, that is, objects $i, i = 1, \ldots, n$ and attributes $j, j = 1, \ldots, m$. The aim of two-mode techniques is to classify the objects and simultaneously to classify the attributes. The main objective of the paper is to present a theoretical data model for two-mode clustering which has the potential to offer a general rationale for the generation of artificial data sets in Monte Carlo experiments.

A great number of different methods has been introduced in the realm of two-mode clustering (cf. Van Mechelen et al. 2004 for an overview). Some authors subdivide these methods into the following three categories (Eckes and Orlik 1993; Schwaiger 1997). The first category consists of methods which are generalizations of the ADCLUS model proposed by Shepard and Arabie (1979; see DeSarbo 1982). The basic assumption is the following: The similarity M_{ij} of an object i and an attribute j is an additive function of weights $V_{kk'}$ which are associated to those

S. Krolak-Schwerdt(✉)
Department of Psychology, Saarland University, Saarbrucken, Germany,
E-mail: s.krolak@mx.uni-saarland.de

A. Okada et al. (eds.), *Cooperation in Classification and Data Analysis*, Studies in Classification, Data Analysis, and Knowledge Organizations.
DOI: 10.1007/978-3-642-00668-5_4,

clusters k and k' the two entities jointly belong to. DeSarbo (1982) developed the first method within this category and Baier et al. (1997) introduced a probabilistic formulation of the model. In general, the input data are required to represent non-symmetric similarities and to be interval scaled. The clustering solution provides overlapping clusters.

The second category contains methods which are fitting additive or ultrametric tree structures to two-mode data. From the two-way input data, a sort of "grand matrix" (with objects and attributes as rows as well as columns) is constructed from which an ultrametric tree structure is estimated. These are hierarchical methods generating non-overlapping solutions. ESOCLUS developed by Schwaiger (1997; cf. Rix and Schwaiger 2003) and the centroid effect method (Eckes and Orlik 1993) belong to this category.

Methods of the third category may be termed "reordering approaches". They perform a reorganization of the data matrix by permuting rows and columns in order to make clusters visible as blocks within the data matrix with objects showing identical values across the attributes. In contrast to the other groups of methods, two-mode profile data are assumed as input which are interpreted as categorical and the algorithms operate directly on the input data without the construction of a grand matrix. Example methods within this category are Hartigan's (1975) two-way joining, the modal block method (Hartigan 1975) or GRIDPAT (Krolak-Schwerdt et al. 1994).

When Monte Carlo studies are conducted to compare the ability of different methods to recover clusters, simulation data derived from a unified model are needed which permits to control and vary features of the cluster structures by certain parameters of that model. Furthermore, the clusters implemented in the simulated data should be within the range of those structures which the two-mode methods are able to detect as clusters.

In our approach, the data are constructed according to the generalized ADCLUS model. The model fulfills the first requirement, as shape parameters of clusters and sizes of the entries of the data are disentangled into three matrices which can be varied for the synthesis of artificial data. Second, the clusters of those data are formed by configurations of blocks of rather homogeneous entries which ought to be recovered not only by methods using ADCLUS, but also by the other methods: (a) reordering methods, as they search for blocks, too, by combinatorial procedures, and (b) methods fitting ultrametric tree structures, as blocks of homogeneous entries may also be interpreted as similarities between objects and attributes.

2 The ADCLUS Model as a General Model for Generating Clustered Data

In our approach, clusters are modelled as patterns of blocks, that is, submatrices of the data matrix with adjacent rows and columns and a certain similarity of the entries within the blocks. Data of this type can be generated in a large variety according to a two-mode generalization of the ADCLUS model[1] which composes a data matrix M as

$$M = RVC' + U + E, \tag{1}$$

where V is of order $K \times K$. R and C are binary matrices of order $n \times K$ and $m \times K$. A row of R corresponds to an object, and a row of C to an attribute. A column of R corresponds to a cluster of objects. Analogously, columns of C correspond to attribute clusters. A two-mode cluster[2] is the union of object cluster k and attribute cluster k' with $V_{kk'} > 0$ (see below).

U is a $n \times m$ matrix with constant entries representing a baseline level of the entries of M. E is a matrix with random entries, each with expectation zero.

The pattern of data generated by RVC' is then represented by a number of blocks which are given by outer products of the column vectors of R and column vectors of C, weighted by appropriate entries of V. More specifically, RVC' consists of blocks of outer products of the following form: k-th column of $R \times k'$th row of C', weighted by $V_{kk'}$. When all entries of V are positive, every combination of an object cluster with an attribute cluster exists. When V is diagonal, every object cluster is combined with exactly one attribute cluster.

The model formulated in (1) is used as a model of data generation not only for procedures which use ADCLUS as a model for the analysis, too, but also for procedures fitting an ultrametric tree and for reordering methods. For the latter, the appropriateness of ADCLUS-like generated data becomes evident when one bears in mind that reordering methods try to organize the data matrix as a pattern of neighboring blocks including nested blocks (see also Van Mechelen et al. 2004 for related arguments).

For the ultrametric tree fitting methods, the appropriateness of ADCLUS-like data is not that obvious. We outline this for the centroid effect method (Eckes and Orlik 1993, abbreviated CEM in the following) and the ESOCLUS algorithm (Schwaiger 1997) as two prominent procedures of this type.

CEM uses a measure of heterogeneity of a subset of the data matrix, which is equal to the variance of the entries in this subset plus the squared difference between the mean of these entries and the overall maximum of the entire data matrix. Due to the fact that the entries within the blocks generated by ADCLUS are of the same size, (a) the variance within blocks is low as compared to any other subsets of entries located across the block structure, and (b) the mean of entries for any subsets within blocks is about the same, so that blocks of ADCLUS-like data correspond to two-mode clusters identified by the centroid effect method.

---

[1] Matrix multiplication AB of two conformable matrices A and B can be looked at from two different perspectives: The entry at position ij of AB is equal to the inner product of the ith row of A and the jth column of B, or AB is the sum of outer products of columns $A_{\cdot i}$ and rows $B_{i \cdot}$, i running through the set of column (row) indices of $A(B)$. When three matrices A, B and C are to be multiplied, the product ABC can be conceived according to the second perspective as the sum of the outer products of columns $A_{\cdot i}$ and rows $C_{j \cdot}$ multiplied by B_{ij}. In our presentation of the ADCLUS model we draw upon the latter point of view.

[2] For a more thorough discussion of the definition and nature of two-mode clusters see Van Mechelen et al. (2005).

In ESOCLUS, a data matrix is considered as a matrix of similarities between objects and attributes. In a first step, the similarities between objects i and attributes j are transformed into distances D_{ij} by

$$D_{ij} = \max_{i,j} M_{ij} - M_{ij} \tag{2}$$

or, if we define $\tilde{M}$ as the constant matrix with all entries equal to $\max_{i,j} M_{i,j}$

$$D = \tilde{M} - M. \tag{3}$$

As it is easily seen, this matrix of distances is again an ADCLUS matrix with the same block structure as the original data. When the design of a simulation study (Krolak-Schwerdt and Wiedenbeck 2006, see below) defines the sizes of entries of the diagonal blocks uniformly larger than all the entries in the off-diagonal blocks with the same rows or columns, then it is evident that the entries with maximum size located in a diagonal block are changed into the smallest object-attribute distances. The block with the second largest entries is a block on the diagonal, too, because of the maximum property again, and its entries are transformed into the second smallest distances, and so on. In contrast, the distance between an object of one block and the attribute of another block is – again because of the maximum property of the diagonal (blocks) – quite large, in any case, larger than any object-attribute distance within a block on the diagonal. In summary, applying ESOCLUS to ADCLUS matrices having the maximum property of diagonal blocks, which includes the use of a hierarchical agglomerative clustering method in a second step (cf. Schwaiger, 1997) to analyze D, will identify the blocks on the diagonal as clusters, depending on the degree of overlap of clusters and the size of random overlay.

2.1 Types of Parameters for the Generation of the Simulated Data

In decomposition (1), different types of parameters for the generation of the data M can be identified. The tools for the design of (non-)overlap are the matrices R and C, the sizes of the entries of the data are essentially controlled by the choice of V. Non-overlapping or disjoint clusters are generated when each row sum of R as well as of C is equal to 1. In case of overlapping clusters row sums exceed 1. Figures 1 and 2 give an illustration of (non-)overlapping in terms of RVC' (cf. Van Mechelen et al. 2004).

Blocks with non-adjacent rows and/or columns or blocks nested within other blocks can be modelled by R and C, too. Therefore, we call R and C the matrices of shape parameters of M.

The levels of the sizes of entries within blocks are determined by matrix V. $V_{kk'}$ is the size of entries of the block of RVC', which is defined by column k of R and column k' of C. In case of overlap, values of V will be added up within regions of

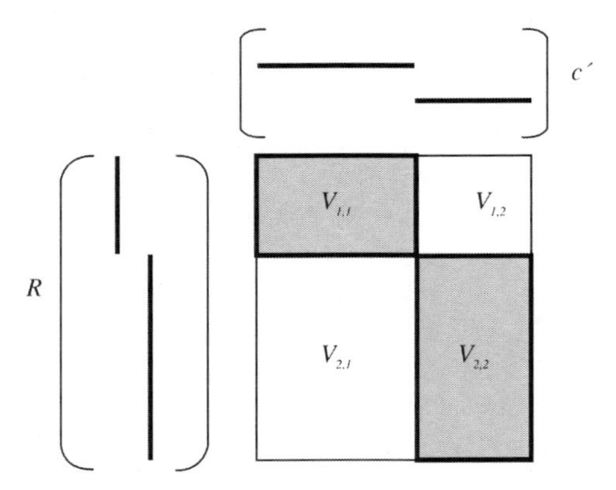

Fig. 1 Two non-overlapping clusters

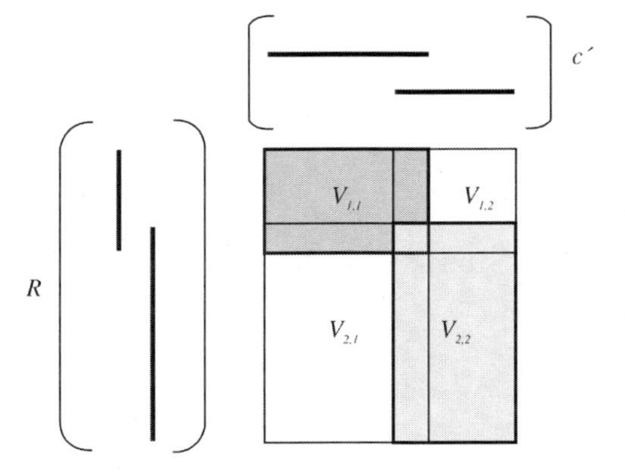

Fig. 2 Two non-overlapping clusters

overlap. The blocks defined by the k-th column of R and the k-th column of C are called "diagonal blocks" of M; its entries are given by the values V_{kk}.

The expectation of all entries of random matrix E is set to zero. The variance, however, can be chosen in various ways, especially different for different parts of M. Finally, the constant matrix U is chosen in a way that M is positive.

2.2 Generating Factors in the Design of Simulated Data M

By specifying the above parameters of M, a large variety of simulated data can be generated. Decisions have to be made concerning the number of blocks of M. Furthermore, the type of relation between blocks has to be specified. That is, should blocks be nested, overlapping (with respect to rows and/or columns), or non-overlapping? Are blocks designed as balanced or unbalanced with respect to the number of object and/or attributes? And how should these different options be combined – should they be varied independently or should the structural pattern of M be designed with compound features, for example combining balancedness with non-overlap or unbalancedness with overlap and/or nesting?

In addition to the structural pattern of blocks, the composition of the size levels are to be determined by the choice of V. For example, which blocks should be given exceptional high values to make them identifiable by clustering methods? Another decision has to be made concerning the relation of the different sizes of the entries. If some values are quite large for some blocks, which blocks should be given smaller values and what should be the difference?

Finally, the identifiability of clusters is influenced by the error matrix E. Here a lot of different options are possible, too. As to the size of the error entries, should they be large compared to the entries of RVC' or should they be rather small? Should the variance of the error terms be constant for all positions M_{ij} or should there be variation, for example depending on the sizes of the blocks of M?

It is obvious, that even for matrices M with a moderate number of objects and attributes, the number of combinations of different patterns is very large. For a simulation study, usually one is forced to limit the ways of variation.

3 An Exemplifying Simulation Study

As an example of the usefulness of the proposed approach, we present in the following a Monte Carlo study[3] conducted by Krolak-Schwerdt and Wiedenbeck (2006) in which some of the possible parameters were varied. The simulated data were analyzed by clustering methods from the class of the generalized ADCLUS models itself (that is, the Baier et al. approach), by methods of fitting an ultrametric tree (these were CEM and ESOCLUS) and by reordering methods (that is, two-way joining and GRIDPAT). In Experiment 1 only data with disjoint clusters were generated, and Experiment 2 data with overlapping clusters.

For both experiments, according to the limitations of the capacity of the applied clustering software, the number of objects was limited to 50 objects and the number of attributes to 20 attributes. Furthermore, the number of columns of R and the number of columns of C were chosen to be equal. Moreover, the main diagonal of V was chosen with a maximum property: For all k, V_{kk} was larger than every remaining entry in the k-th row as well as in the k-th column of V.

Finally, a random matrix E was superimposed. For Experiment 1, the entries of E were uniformly distributed, and, for Experiment 2, normally distributed with expectation zero, respectively.

For Experiment 1, clusters were defined at varying conditions or factors which influence the *separability* of the clusters. Two factors were chosen (1) Number of clusters: Data sets with three, five and eight clusters were generated. (2) For each

[3] Only variations of parameters of major importance are reported. For the full variations of parameters within this study, see Krolak-Schwerdt and Wiedenbeck (2006).

number of clusters the shapes of clusters were varied by generating a balanced or an unbalanced number of entities (objects and/or attributes) per cluster. This factor is called "structure" and had four levels: Structure 1 comprised a balanced number of objects as well as of attributes. Structure 2 was balanced in the number of objects and unbalanced in the number of attributes. Structure 3 had unbalanced numbers of objects and balanced numbers of attributes. In structure 4, high numbers of objects were combined with low numbers of attributes, and vice versa.

Standard deviations varied between clusters (in case of three clusters, values are chosen as 2, 3, and 4).

For Experiment 1, the Adjusted Rand Index (ARI) of Hubert and Arabie (1985, p. 198) was chosen as an indicator of the solution recovery quality.

We performed analysis of variance (ANOVA) on the ARI indices with the number of clusters (3, 5 or 8), type of structure (structure 1, 2, 3 or 4) and type of method (Baier et al. approach, CEM, ESOCLUS, two-way joining or GRIDPAT) as experimental factors.

The basic data pattern of Experiment 2 was extended in that maximally two clusters were allowed to overlap with respect to objects or attributes or both. The basic pattern of V was constructed using Toeplitz matrices of the following type:

$$\begin{pmatrix} D_1 & \lambda D_1 & \lambda^2 D_1 & \lambda^3 D_1 & \cdots \lambda^{K-1} D_1 \\ \lambda D_1 & D_2 & \lambda D_2 & \lambda^2 D_2 & \cdots \lambda^{K-2} D_2 \\ \lambda^2 D_1 & \lambda D_2 & D_3 & \lambda D_3 & \cdots \lambda^{K-3} D_3 \\ \lambda^3 D_1 & \lambda^2 D_2 & \lambda D_3 & D_4 & \cdots \lambda^{K-4} D_4 \\ \cdots & & & & \\ \lambda^{K-1} D_1 & \lambda^{K-2} D_2 & \lambda^{K-3} D_3 & \lambda^{K-4} D_4 \cdots & D_K \end{pmatrix}.$$

For $0 < \lambda < 1$ and $D_1 > 0$, the entry D_1 at the uppermost left position on the diagonal is the maximum element of the first row and first column. By cancellation of the first row and first column, a $(K-1) \times (K-1)$-matrix of the same type is left. Again, an analogous maximum property is maintained for a positive real entry D_2 at the uppermost left end of the diagonal in the residual matrix. Continuing so with positive numbers $D_3, \ldots, D_K$ and reimplementing the cancelled rows and columns, a matrix V is generated with the above mentioned maximum property of diagonal entries. In this type of matrix, the entries on the diagonal are the maximum entries of their corresponding rows and columns, if $\lambda D_k \leq D_{k+1}, k = 1, \ldots, K-1$, which is given for our data.

For Experiment 2, one Toeplitz matrix with $\lambda = 0.6$ and one with $\lambda = 0.2$ was used. Obviously, the smaller λ favours the separability of the clusters. The numbers D_k were chosen to be 20, 30, 40 and so on, depending on the number of clusters.

For Experiment 2, the $Omega$ index of Collins and Dent (1988) was selected. $Omega$ is a generalization of ARI to the case of overlapping clustering solutions.

Analysis of variance was performed on the $Omega$ indices with the number of clusters (3, 5, or 8), degree of cluster overlap (large or small), magnitude of parameter in the Toeplitz matrix (0.2 or 0.6) and type of method (Baier et al. approach, centroid effect method, ESOCLUS, two-way joining or GRIDPAT) as experimental factors.

3.1 Results

First, the results of Experiment 1 are reported. As to the central aim of the Monte Carlo study, in Experiment 1 the main effect of methods, $F(4, 180) = 56.66$, $p < 0.001$, is of importance. Across all data sets, ESOCLUS ($\bar{x} = 0.81$) performed best followed by CEM ($\bar{x} = 0.81$), Hartigan's two-way joining ($\bar{x} = 0.69$), GRIDPAT ($\bar{x} = 0.66$) and the Baier et al. approach performing poorest ($\bar{x} = 0.52$). However, the main effect of methods was accentuated by two interactions.

The first is a strong interaction of cluster number and method, $F(8, 180) = 68.44$, $p < 0.001$. Table 1a shows the mean ARI indices as a function of the corresponding factor level combinations. In this experiment as well as in Experiment 2 standard errors were in the range between 0.00 and 0.07 and thus were low. ESOCLUS, CEM and GRIDPAT decrease in recovery with increasing cluster number. However, Hartigan's two-way joining shows the reverse behavior, in that recovery quality is rather low in three and five cluster data sets, but dramatically increases in the eight cluster configurations. Finally, the Baier et al. procedure is unaffected by the cluster number on a comparatively low level of cluster recovery.

The second interaction[4] involves cluster structure and method, $F(12, 180) = 3.12$, $p < 0.001$. From the balanced structure 1 to the most unbalanced structure 4, ESOCLUS, CEM and the Baier et al. approach display a monotonic decrease in performance, while GRIDPAT and Hartigan's two-way joining remain rather constant.

In the following, results of Experiment 2 are presented. The significant interaction of cluster number and method, $F(8, 240) = 25.25$, $p < 0.001$, was replicated in Experiment 2. However, the interaction was accentuated by the triple interaction of cluster number, method and degree of cluster overlap, $F(8, 240) = 6.61$, $p < 0.001$.

Table 1 Mean values of ARI from Experiment 1 and of $Omega$ from Experiment 2 as a function of method, cluster number and cluster overlap

	Experiment 1			Experiment 2					
	1(a) No overlap – disjoint			1(b) Large overlap			1(c) Small overlap		
Number of clusters	3	5	8	3	5	8	3	5	8
Method									
ESOCLUS	0.99	0.90	0.52	0.80	0.81	0.81	0.88	0.90	0.91
CEM	0.96	0.78	0.57	0.81	0.81	0.82	0.87	0.91	0.92
Baier et al.	0.54	0.53	0.50	0.92	0.85	0.78	0.85	0.84	0.86
GRIDPAT	0.90	0.76	0.30	0.92	0.87	0.82	0.82	0.87	0.87
Two-way joining	0.63	0.48	0.94	0.80	0.89	0.76	0.68	0.78	0.79

[4] Due to space limitations the reader may be referred to Krolak-Schwerdt and Wiedenbeck (2006) for the corresponding values of ARI.

Tables 1b and 1c show the mean *Omega* indices as a function of the corresponding factor level combinations.

There is one group of methods consisting of ESOCLUS and CEM which exhibits rather constant recovery values across all levels of cluster number. These recovery values are high in the presence of small clustering overlap, but decrease considerably in the data sets involving large overlap.

The Baier et al. approach, Hartigan's two-way joining and GRIDPAT perform differently depending on cluster number and cluster overlap. For a smaller number of clusters, these methods exhibit high recovery values in the presence of large cluster overlap, but perform poorer in case of small overlap. Furthermore, recovery values decrease with increasing cluster number if cluster overlap is large, while the *Omega* indices remain rather constant across cluster numbers in data sets with small overlap.

Finally, as compared to ESOCLUS and CEM, the recovery of methods of the second group is poorer for clusters with small overlap. Altogether, then, the Baier et al. approach, two-way joining and GRIDPAT outperform ESOCLUS and CEM in the analysis of data structures with a smaller number of highly overlapping clusters. In the presence of only a small degree of cluster overlap, the results are similar to those of Experiment 1 in that the non-overlapping methods (e.g., ESOCLUS and CEM) are superior to the second group of methods in recovering the true clustering structures.

4 Conclusions

For our simulation study a data model was introduced which may offer the potential for an integration of methods. The proposed model has close relationships to the two-mode formulation of the ADCLUS model. The ADCLUS framework establishes the mathematical link between ADCLUS generalizations and reordering methods (cf. Van Mechelen et al. 2004) and so does the data model introduced in this paper. Additionally, it allows for a restriction of parameters such that the type of data underlying the non-overlapping hierarchical methods is derived. Thus, the proposed model of two-mode data serves as a common frame of reference where parameter restrictions generate data with the properties of one or the other group of methods.

Future research will show if the proposed model may serve as a frame of reference to integrate methods from the different categories. If this holds true, than the following theoretical underpinning is associated with the proposed approach (see also DeSarbo 1982): The model assumes that people arrive at a judgment of the similarity between entities such as an object and an attribute in an additive way. The basis for the similarity judgment are features or properties which underlie the entities. These features are represented by the clusters. Summing the weights of those properties or features the corresponding two entities have in common represents the process and the outcome of the judgment.

What is not known at the current stage of the investigation is in how far other two-mode methods than the ones mentioned in this paper may be integrated in the proposed framework, too. As an example, De Soete et al. (1984) introduced a method of fitting additive trees to two-mode data with a different rationale than CEM and ESOCLUS. For now, the possibility to subsume this approach and many others developed to analyze two-mode data remains an open question.

The results on the recovery performance of the two-mode methods revisited in this paper suggest that the ability of the methods to recover a given clustering structure strongly depends on the type and complexity of the data structure. Most notably, methods of each category performed best if the input data corresponded to the data structure presumed by the method. That is, for ADCLUS generalizations and reordering methods, we found superior performance in the presence of highly overlapping clusters, while methods estimating ultrametric tree structures performed best in analyzing non-overlapping clusters or clusters with only a small degree of overlap.

Consequently, if methods are to be used in a heuristic way due to a lack of some a priori knowledge, the best what could be done is to select one of the hierarchical methods as they performed best across all conditions. However, as our Monte Carlo results have shown, this might not guarantee an optimal solution in every case.

References

BAIER, D., GAUL, W. and SCHADER, M. (1997): Two-mode overlapping clustering with applications to simultaneous benefit segmentation and market structuring. In: R. Klar and O. Opitz (Eds.): *Classification and Knowledge Organization*. Springer, Berlin, 557–566.

COLLINS, L.M. and DENT, C.W. (1988): Omega: A general formulation of the Rand index of cluster recovery suitable for non-disjoint solutions. *Multivariate Behavioral Research, 23,* 231–342.

DE SOETE, G., DESARBO, W.S., FURNAS, G.W. and CARROLL, D.J. (1984): The estimation of ultrametric and path length trees from rectangular proximity data. *Psychometrika, 49,* 289–310.

DESARBO, W.S. (1982): Gennclus: New models for general nonhierarchical clustering analysis. *Psychometrika, 47,* 449–475.

ECKES, T. and ORLIK, P. (1993): An error variance approach to two-mode hierarchical clustering. *Journal of Classification, 10,* 51–74.

HARTIGAN, J. (1975): *Clustering Algorithms*. Wiley, New York.

HUBERT, L. and ARABIE, P. (1985): Comparing partitions. *Journal of Classification, 2,* 193–218.

KROLAK-SCHWERDT, S. and WIEDENBECK, M. (2006): The recovery performance of two-mode clustering methods: A Monte Carlo experiment. In M. Spiliopoulou, R. Kruse, C. Borgelt, A. Nürnberger and W. Gaul (Eds.): *Studies in Classification, Data Analysis and Knowledge Organization, Vol. 29: From Data and Information Analysis to Knowledge Engineering*. Springer, Berlin, 190–197.

KROLAK-SCHWERDT, S., ORLIK, P. and GANTER, B. (1994): TRIPAT: a model for analyzing three-mode binary data. In: H.H. Bock, W. Lenski and M.M. Richter (Eds.): *Studies in Classification, Data Analysis, and Knowledge Organization, Vol. 4 Information systems and data analysis*. Springer, Berlin, 298–307.

RIX, R. and SCHWAIGER, M. (2003): Two-mode hierarchical cluster analysis – evaluation of goodness-of-fit measures. *Münchner Betriebswirtschaftliche Beiträge* 5(2003).

SCHWAIGER, M. (1997): Two-mode classification in advertising research. In: R. Klar and O. Opitz (Eds.): *Classification and Knowledge Organization*. Springer, Berlin, 596–603.

SHEPARD, R.N. and ARABIE, P. (1979): Additive clustering representation of similarities as combinations of discrete overlapping properties. *Psychological Review, 86,* 87–123.

VAN MECHELEN, I., BOCK, H. and DE BOECK, P. (2004): Two-mode clustering methods: A structure overview. *Statistical Methods in Medical Research, 13,* 363–394.

VAN MECHELEN, I., BOCK, H. and DE BOECK, P. (2005): Two-mode clustering. In: B.S. Everitt and D.C. Howell (Eds.): *Encyclopedia of Statistics in Behavioral Science, Vol. 4*, Wiley, New York, 2081–2086.

Density-Based Multidimensional Scaling

F. Rehm, F. Klawonn, and R. Kruse

Abstract Multidimensional scaling provides dimensionality reduction for high-dimensional data. Most of the available techniques try to preserve similarity in terms of distances between data objects. In this paper a new approach is proposed that extends the distance preserving aspect by means of density preservation. Combining both, the distance aspect and the density aspect, permits efficient multidimensional scaling solutions.

1 Introduction

Most branches of commerce, industry and research put great efforts in collecting data with the objective to describe and predict customer behaviour or both technical and natural phenomena. Besides the size of such data sets, data analysis becomes challenging due to a large number of attributes describing a data object. Visualization can facilitate the discovery of structures, patterns and relationships in data and exploratory visualization is an important component in hypothesis generation.

Multidimensional scaling (MDS) is a family of dimensionality reduction techniques that use optimization to preserve distance relationships between points in the multidimensional space in the two- or three-dimensional mapping required for effective visualization (Kruskal and Wish 1978). In the recent years much effort has been done to improve MDS regarding its computational complexity (Borg and Groenen 2005; Chalmers 1996; Morrison et al. 2003; Williams and Munzner 2004). Besides distance-based approaches also some techniques preserving angles between data objects have been applied successfully (Lesot et al. 2006; Rehm et al. 2006).

In this paper we present a new approach that extends conventional distance-based multidimensional scaling by a density preserving aspect. This permits to improve

F. Rehm(✉)
Institute of Flight Guidance, German Aerospace Center, Braunschweig, Germany,
E-mail: frank.rehm@dlr.de

A. Okada et al. (eds.), *Cooperation in Classification and Data Analysis*,
Studies in Classification, Data Analysis, and Knowledge Organizations.
DOI: 10.1007/978-3-642-00668-5_5,

the mapping of high-dimensional data for visualization purposes. The rest of the paper is organized as follows. In Sect. 2 we briefly review Sammon's mapping as a common representative of distance-based MDS. Section 3 describes the proposed method. Section 4 discusses results on benchmark examples. Finally we conclude with Sect. 5.

2 Sammon's Mapping

Sammon's mapping is a multidimensional scaling technique that estimates the coordinates of a set of objects $Y = \{y_1, \ldots, y_n\}$ in a feature space of specified (low) dimensionality that come from data $X = \{x_1, \ldots, x_n\} \subset \mathbb{R}^p$ trying to preserve the distances between pairs of objects. These distances are usually stored in a distance matrix

$$D^x = \left(d_{ij}^x\right), \quad d_{ij}^x = \left\|x_i - x_j\right\|, \quad i, j = 1, \ldots, n.$$

The estimation of the coordinates will be carried out under the constraint that the error between the distance matrix D^x of the data set and the distance matrix $D^y = (d_{ij}^y)$, $d_{ij}^y = \|y_i - y_j\|$, $i, j = 1, \ldots, n$ of the corresponding transformed data set will be minimized.

Different error measures to be minimized were proposed, e.g., the absolute error that considers non-weighted differences between original distances and distances in the target space, the relative error that takes relative distances into account or a combination of both. The Sammon's mapping error measure

$$E_{\text{sammon}} = \frac{1}{\sum_{i=1}^{n} \sum_{j=i+1}^{n} d_{ij}^x} \sum_{i=1}^{n} \sum_{j=i+1}^{n} \frac{\left(d_{ij}^y - d_{ij}^x\right)^2}{d_{ij}^x} \tag{1}$$

describes the absolute and the relative quadratic error. To determine the transformed data set Y by means of minimizing error E_{sammon} a gradient descent method can be used. By means of this iterative method, the parameters y_l to be optimized, will be updated during each step proportional to the gradient of the error function E. Calculating the gradient of the error function leads to

$$\frac{\partial E_{\text{sammon}}}{\partial y_l} = \frac{2}{\sum_{i=1}^{n} \sum_{j=i+1}^{n} d_{ij}^x} \sum_{j \neq l} \frac{d_{lj}^y - d_{lj}^x}{d_{lj}^x} \frac{y_l - y_j}{d_{lj}^y}. \tag{2}$$

After random initialization for each projected feature vector y_l a gradient descent is carried out and the distances d_{ij}^y as well as the gradients $(\partial d_{ij}^y / \partial y_l)$ will be recalculated again. The algorithm terminates when E_{sammon} becomes smaller than a certain threshold.

3 Density-Based Mappings

The concept of density-based visualization is to map density distributions of high-dimensional data into low-dimensional feature spaces. Density variations often indicate the existence of clusters which, commonly, are of concern in the field of data mining. Thus, projecting these density distributions to a visually interpretable display may help to identify interesting patterns in the data.

In the following we formalize the problem of density preservation by means of an objective function that can be minimized through a gradient descent technique. For each data object in the original data space a multivariate Gaussian distribution is defined that represents a data point's potential energy. When adding those single potentials we get a sort of multidimensional potential mountains. Summits of the mountains can be found where many data objects are located. Accordingly, valleys can be found in areas of low data density.

Similarly, one can reproduce the mountains in the low-dimensional feature space (usually two or three dimensions). For this purpose each data object of the original space will be placed in the projection space. Over every single data point a potential (in form of a two- or three-dimensional Gaussian distribution) will be applied. The criterion for the mapping is that the potentials in the original space coincide as good as possible with the potentials at the corresponding points in the target space.

Given the data set $X = \{x_1, \ldots, x_n\} \subset \mathbb{R}^p$ we seek for the mapped data set $Y = \{y_1, \ldots, y_n\} \subset \mathbb{R}^k$ with $k = 2$ or $k = 3$ with the following potential for x_i:

$$f_i(x) = \frac{1}{c} \exp\left[-\frac{1}{2} \sum_{t=1}^{p} \left(\frac{x^{(t)} - x_i^{(t)}}{\sigma} \right)^2 \right] \tag{3}$$

with

$$c = \frac{1}{\sigma^p \sqrt{(2\pi)^p}}.$$

By $x^{(t)}$ and $x_i^{(t)}$ we denote the t-th attribute of data object x and x_i, respectively. Function f_i simply describes the density of a p-dimensional Gaussian distribution with mean value x_i and variance σ^2 in each dimension. The parameter σ must be fixed for the entire procedure. If σ is rather small, then the potentials do rarely overlap. For very large σ the potential landscape will be blurred completely with little variance in height.

Therefore, it is useful to define σ according to the diameter d of the data space, the average distance between data points, the number n of data objects and the dimensionality p. A straight forward approach would be to assume that the data is uniformly distributed in a hyper-cube or hyper-sphere. In this case the potentials would have approximately the same height. Of course, this assumption is fairly theoretical. In practice mountains will be formed due to the heterogeneous structure of the data. However, under this assumption the average density can be computed and the potentials on and between data points can be determined. The larger the variance σ^2, the smaller the difference in the potentials. For small data sets the density is low and therefore a larger σ should be chosen.

Similar to Sammon's mapping we seek the projected data points $Y = \{y_1, \ldots, y_n\} \subset \mathbb{R}^k$. Over each data point we apply a potential (in this case a k-dimensional Gaussian distribution) as for the original space:

$$g_i(y) = \frac{1}{\tilde{c}} \exp\left[-\frac{1}{2} \sum_{t=1}^{k} \left(\frac{y^{(t)} - y_i^{(t)}}{\tilde{\sigma}} \right)^2 \right] \tag{4}$$

with

$$\tilde{c} = \frac{1}{\tilde{\sigma}^k \sqrt{(2\pi)^k}}.$$

Then the objective is to place the feature vectors such that the potentials coincide at least in these points with those in the original space. Note, $\tilde{\sigma}$ should be chosen similarly to σ. In the ideal case we have approximately the same diameter d in the target space, too. However, the area (or the volume) of the target space will be much smaller compared to the hyper volume of the original space ($k \ll p$). This means that the density in the target space is also higher for the same size of the data set. Thus, $\tilde{\sigma}$ should be chosen smaller than σ. Still the potentials in the target space might not match the potentials in the original space yet. It should be assured that the maximum height of the single potentials in the original space and in the target space match, i.e., the respective maxima of the Gaussian distributions should be:

$$f_i(x_i) \approx g_i(y_i).$$

Since normally this will not be the case we introduce a constant a:

$$a f_i(x_i) = g_i(y_i)$$

which can be derived from (3) and (4):

$$a = \frac{\sigma^p}{\tilde{\sigma}^k} \sqrt{(2\pi)^{p-k}}.$$

Now we can formulate our objective function. The summarized modified potential in the original space at x_i is

$$\sum_{j=1}^{n} a f_j(x_i)$$

and in the target space at y_i

$$\sum_{j=1}^{n} g_j(y_i).$$

In the ideal case, both potentials should be equal. Hence, we define the objective function as follows:

$$E_{\text{density}} = \sum_{i=1}^{n} \left[\sum_{j=1}^{n} g_j(y_i) - \sum_{j=1}^{n} af_j(x_i) \right]^2 = \sum_{i=1}^{n} \left\{ \sum_{j=1}^{n} \left[g_j(y_i) - af_j(x_i) \right] \right\}^2. \quad (5)$$

Now, we only have to determine the gradient for each component s:

$$\frac{\partial E_{\text{density}}}{\partial y_{ls}} = 2 \sum_{i=1}^{n} \sum_{j=1}^{n} \left[g_j(y_i) - af_j(x_i) \right] \frac{\partial}{\partial y_{ls}} g_j(y_i). \quad (6)$$

$\frac{\partial}{\partial y_{ls}} g_j(y_i)$ is only zero when we have $l = i$ or $l = j$. For both cases we derive from (6):

$$\frac{\partial}{\partial y_{ls}} g_l(y_i) = \frac{1}{\tilde{c}} \exp\left[-\frac{1}{2} \sum_{t=1}^{k} \left(\frac{y_i^{(t)} - y_l^{(t)}}{\tilde{\sigma}} \right)^2 \right] \frac{y_i^{(s)} - y_l^{(s)}}{\tilde{\sigma}}, \quad (7)$$

$$\frac{\partial}{\partial y_{ls}} g_j(y_l) = -\frac{1}{\tilde{c}} \exp\left[-\frac{1}{2} \sum_{t=1}^{k} \left(\frac{y_l^{(t)} - y_j^{(t)}}{\tilde{\sigma}} \right)^2 \right] \frac{y_l^{(s)} - y_j^{(s)}}{\tilde{\sigma}}. \quad (8)$$

It can be easily seen that for $i = j = l$ we have $\frac{\partial}{\partial y_{ls}} g_j(y_i) = 0$. Finally we obtain for the gradient:

$$\begin{aligned} \frac{\partial E_{\text{density}}}{\partial y_{ls}} = \frac{2}{\tilde{c}} \sum_{i=1}^{n} \Bigg\{ & \left[g_l(y_i) - af_l(x_i) \right] \\ & \times \exp\left[-\frac{1}{2} \sum_{t=1}^{k} \left(\frac{y_i^{(t)} - y_l^{(t)}}{\tilde{\sigma}} \right)^2 \right] \frac{y_i^{(s)} - y_l^{(s)}}{\tilde{\sigma}} \\ & - \left[g_i(y_l) - af_i(x_l) \right] \\ & \times \exp\left[-\frac{1}{2} \sum_{t=1}^{k} \left(\frac{y_l^{(t)} - y_i^{(t)}}{\tilde{\sigma}} \right)^2 \right] \frac{y_l^{(s)} - y_i^{(s)}}{\tilde{\sigma}} \Bigg\}. \end{aligned} \quad (9)$$

Combining the Sammon gradient E_{sammon} and the density gradient E_{density} through linear combination we finally obtain

$$E = \alpha \frac{\partial E_{\text{sammon}}}{\partial y_l} + \beta \frac{\partial E_{\text{density}}}{\partial y_l}. \quad (10)$$

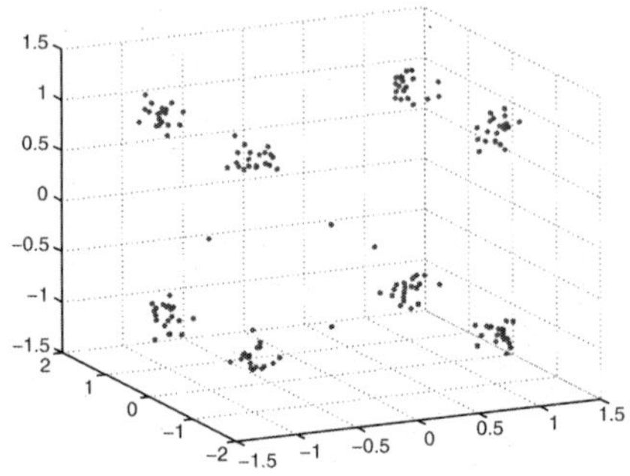

Fig. 1 Cube data set

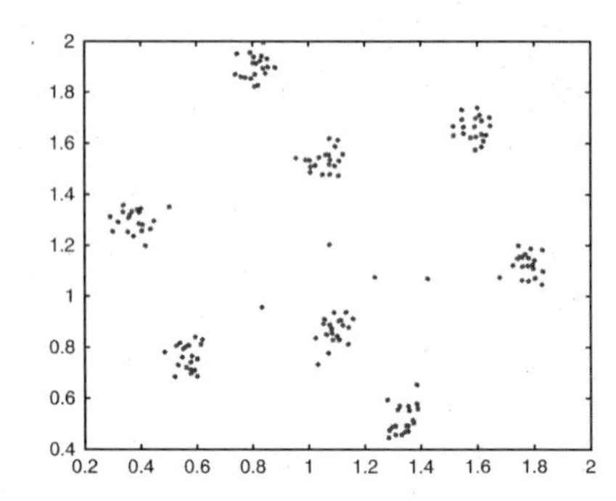

Fig. 2 Sammon's mapping of the Cube data set

The parameters $\alpha \geq 0$ and $\beta \geq 0$ can be considered as learning rates or weights to control the impact of the respective mapping strategy. Thus, higher weights α for the Sammon gradient favour distance-based mappings and larger values β for the density gradient favour the density approach.

4 Results

In this section we will discuss some results of the proposed technique on some benchmark examples. The first data set, the Cube data set (see Fig. 1), is about a synthetic data set, where data points scatter around the corners of an imaginary three-dimensional cube. Thus, the Cube data set contains eight well separated clusters. The second data set, the Wine data set (Forina et al. 1988), results from a chemical analysis of wines grown in the same region in Italy but derived from three different cultivars. The analysis determined the quantities of 13 constituents found in each of the three types of wines.

Figure 2 shows a Sammon's mapping of the Cube data set. The eight data clusters are well reflected in the mapping. The transformation with the density-based approach, setting $\alpha = 0$ and thusly optimizing the density aspect exclusively, leads to the mapping visualized in Fig. 3. It is surprising that already the density aspect in the optimization is sufficient in this example to reflect the structure of the data set. Applying a linear combination of both, the Sammon gradient and the density

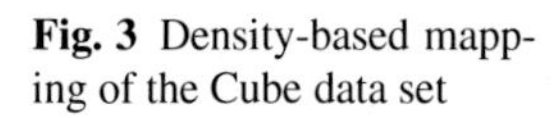

Fig. 3 Density-based mapping of the Cube data set

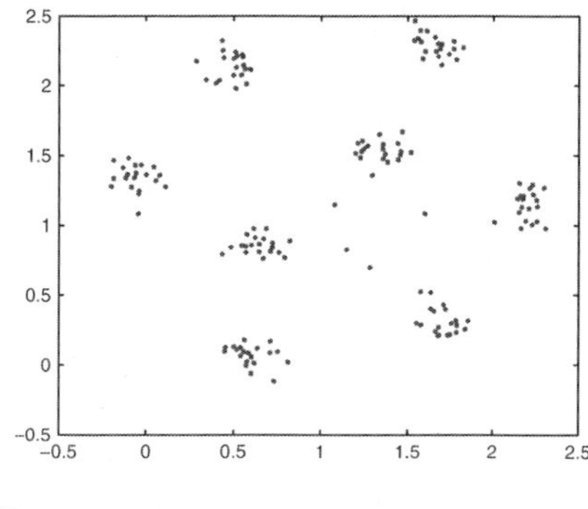

Fig. 4 Mapping of the Cube data set (distance-based and density-based)

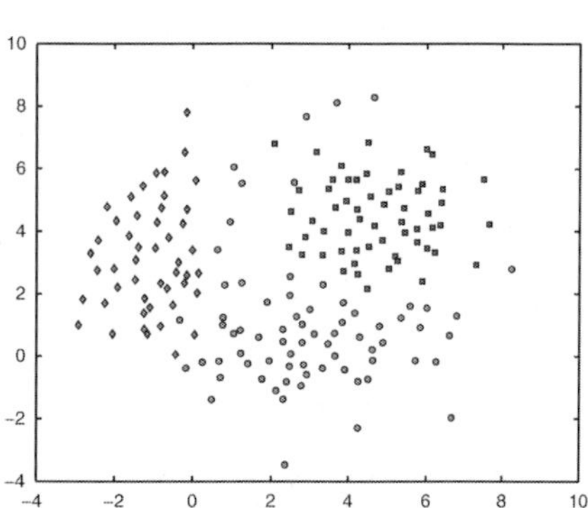

Fig. 5 Sammon's mapping of the Wine data set

gradient, we obtain the mapping depicted in Fig. 4. Whereas the distance-based approach seams to favour the preservation of the inter-cluster structure, the linear combination of distance and density aspects gives a better overall impression of the data set.

Figures 5 and 6 show transformations of the Wine data set with Sammon's mapping and with the density-based approach, respectively. Both transformations show similar characteristics.

Based on the empirical tests we cannot constitute that the density-based approach is superior to the distance-based approach. Indeed, the computational complexity per iteration of the density-based approach is rather higher since the density gradient

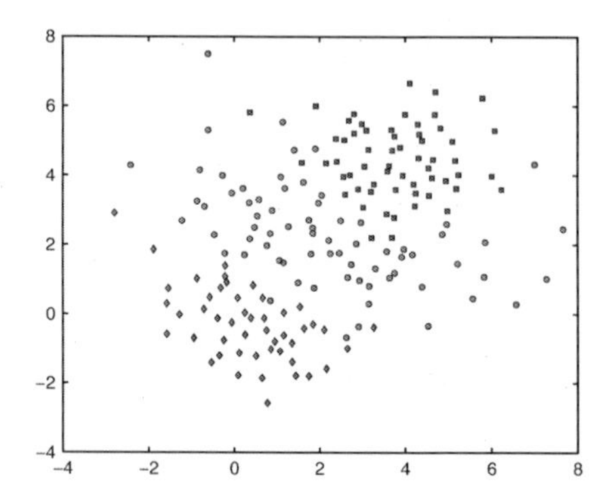

Fig. 6 Density-based mapping of the Wine data set

has to be computed additionally. Our tests have shown that the number of iterations can be reduced with density preservation.

5 Conclusions

In this paper we have presented a new approach to visualize high-dimensional data. Density-based multidimensional scaling considers not only the distance aspect as it is usual but also density aspects of a data set. We could show that our approach is promising and leads to comparable results as conventional MDS and can lead to better results in combination with MDS. Future work should focus on further tests on complex data sets to prove stability and convergence.

References

BORG, I. and GROENEN, P. (2005): *Modern Multidimensional Scaling: Theory and Applications*. Springer, Berlin.

CHALMERS, M. (1996): A Linear Iteration Time Layout Algorithm for Visualising High-Dimensional Data. In *Proceedings of IEEE Visualization 1996*, San Francisco, CA, 127–132.

FORINA, M., LEARDI, R., ARMANINO, C. and LANTERI, S. (1988): *PARVUS: An Extendable Package of Programs for Data Exploration, Classification and Correlation.* Elsevier, Amsterdam.

KRUSKAL, J.B. and WISH, M. (1978): *Multidimensional Scaling.* Sage, Beverly Hills.

LESOT, M.J., REHM, F., KLAWONN, F. and KRUSE, R. (2006): Prediction of Aircraft Flight Duration. In *Proceedings of the 11th IFAC Symposium on Control in Transportation Systems,* Delft, 107–112.

MORRISON, A., ROSS, G. and CHALMERS, M. (2003): Fast Multidimensional Scaling through Sampling, Springs and Interpolation. *Information Visualization*, 2, 68–77.

REHM, F., KLAWONN, F. and KRUSE, R. (2006): POLARMAP – Efficient Visualisation of High Dimensional Data. In *IEEE Proceedings of the 10th International Conference on Information Visualisation*, London, 731–740.

WILLIAMS, M. and MUNZNER, T. (2004): Steerable, Progressive Multidimensional Scaling. In *Proceedings of the 10th IEEE Symposium on Information Visualization,* Austin, TX, 57–64.

Classification of Binary Data Using a Spherical Distribution

Y. Sato

Abstract Recently, in the trend of this field, for instance, data mining and statistical learning theory, the non-parametric methods seem to be a mainstream. The major reason why almost every method are non-parametric, they said that any assumptions are not required for the observation. This is true for the training the model. But for the prediction, this situation seems to be different, because the criterion for fitting the model will be observed data (training data) itself. Furthermore, in these methods, neural network model, support vector machine, the set of the feasible function seems to be too large. In other words, for instance, the degree of freedoms or VC dimensions is almost infinite. This implies the over fitting problem in most cases. A natural or moderate assumption of the distribution for the observation leads to better results for the prediction. The parametric method corresponds to the restriction of the feasible functions. Linear discriminant function is a typical one. Of cause, there is the concept of semi-parametric, for instance, a mixture model and so on. But this will be discussed elsewhere.

1 Introduction

A binary observations are defined as a data whose attributes are qualitative, and each attribute has two state, positive response or negative response. Usually, these responses are denoted by 0 or 1. But their values themselves have no meaning. Then in this paper, we denote 1 or 2 for the sake of the transformation stated below.

A classification problem is intrinsically dependent on a distance structure of the data. When we classify the data into some groups, we have to define a suitable similarity measure among objects. The most simple similarity or dissimilarity is the

Y. Sato
Division of Computer Science, Graduate School of Information Science and Technology, Hokkaido University, Kita 14, Nishi 9, Kita-ku, Sapporo 060-0814, Japan,
E-mail: ysato@main.ist.hokudai.ac.jp

A. Okada et al. (eds.), *Cooperation in Classification and Data Analysis*,
Studies in Classification, Data Analysis, and Knowledge Organizations.
DOI: 10.1007/978-3-642-00668-5_6,

Table 1 Typical similarity measures for binary data

Similarity measures	
Matching coefficient	$\frac{a+d}{p}$
Jaccard coefficient (1908)	$\frac{a}{a+b+c}$
Rogers and Tanimoto (1960)	$\frac{a+d}{a+2(b+c)+d}$
Sokal and Sneath (1963)	$\frac{a}{a+2(b+c)}$
Gower and Legendre (1986)	$\frac{a+d}{a+\frac{1}{2}(b+c)+d}$
	$\frac{a}{a+\frac{1}{2}(b+c)}$

distance relation. The definitions and discussions of the similarity and dissimilarity between a pair of binary vectors have a long history.

When we denote the object i and object j by the following binary vectors,

$$\boldsymbol{x}_i = (x_{i1}, x_{i2}, \cdots, x_{ip}), \quad \boldsymbol{x}_j = (x_{j1}, x_{j2}, \cdots, x_{jp})$$

we count the number of attributes whose pair (x_{ir}, x_{jr}) has the binary values $(2,2), (2,1), (1,2), (1,1)$, and we denote a,b,c,d, respectively. Then the typical similarity measures are denoted in Table 1. In these similarity measures, first we find out that the term a and d are not dealt with equally. That is, whether we count the "negative matching" for similarity or not. This means that if two objects don't have some property in common, we should count this or not. On the other hand, the major difference in these similarities is the weight of the term $(b+c)$. Then we may consider the suitable weight for the term $(b+c)$. But this lead us to the idea that we should weight the vectors $\boldsymbol{x}_i$ themselves in a suitable way. Then in this paper, we weight each vectors $\boldsymbol{x}_i$ by the reciprocal of the sum of components. This is a basic idea of the transformation binary data into directional data.

2 Transformation Binary Data into Directional Data

We assume that the following n binary objects with p attributes are given

$$\boldsymbol{x}_i = (x_{i1}, x_{i2}, \cdots, x_{ip}), \quad x_{ia} = 1 \text{ or } 2, \quad (i = 1, 2, \ldots, n).$$

We suppose that each object $\boldsymbol{x}_i$ is weighted by the sum of the value of attributes, i.e. sum of the component of the vector $\boldsymbol{x}_i$. When we denote the weighted vector as $\boldsymbol{\xi}_i = (\xi_{i1}, \xi_{i2}, \ldots, \xi_{ip})$, the components are given by

$$\xi_{ia} = x_{ia} / \sum_{b=1}^{p} x_{ib}.$$

Since each $\boldsymbol{\xi}_i$ has the following relation,

$$\sum_{a=1}^{p} \xi_{ia} = 1, \ \xi_{ia} > 0, \tag{1}$$

we know that the vectors $\boldsymbol{\xi}_i$ are located as an inner point on $(p-1)$-dimensional hyperplane in the first quadrant of p-dimensional space. We must introduce a suitable metric function on this hyperplane. Since $\boldsymbol{\xi}_i$ has the property in (1), we can use an analogy of a discrete probability distribution, i.e. if we regard $\boldsymbol{\xi}_i$ as a probability, then we are able to introduce Kullback–Leibler divergence as a distance measure, correspondingly, which are defined as follows:

$$D(\boldsymbol{\xi}_i, \boldsymbol{\xi}_j) = \frac{1}{2} \sum_{a=1}^{p} (\xi_{ia} - \xi_{ja}) \log \frac{\xi_{ia}}{\xi_{ja}}.$$

In order to investigate the structure of the space, we may consider a line element between two different points, $\boldsymbol{\xi}_i$ and $\boldsymbol{\xi}_i + d\boldsymbol{\xi}_i$. When we evaluate Kullback–Leibler divergence between these two points up to the second order with respect to $d\boldsymbol{\xi}_i$, we get

$$D(\boldsymbol{\xi}_i + d\boldsymbol{\xi}_i, \boldsymbol{\xi}_i) = \frac{1}{2} \sum_{a=1}^{p} d\xi_{ia} \log \frac{\xi_{ia} + d\xi_{ia}}{\xi_{ja}} \approx \frac{1}{4} \sum_{a=1}^{p} \frac{1}{\xi_{ia}} (d\xi_{ia})^2.$$

This is well known as a chi-square distance. However, since the dimension of this space (hyperplane) is $(p-1)$, when we substitute the following relation into the above,

$$\xi_{ip} = 1 - \sum_{a=1}^{p-1} \xi_{ia}, \quad d\xi_{ip} = - \sum_{a=1}^{p-1} d\xi_{ia}$$

we get

$$D(\boldsymbol{\xi}_i + d\boldsymbol{\xi}_i, \boldsymbol{\xi}_i) = \frac{1}{4} \sum_{a=1}^{p-1} \sum_{b=1}^{p-1} \left(\delta_{ab} \frac{1}{\xi_{ia}} + \frac{1}{\xi_{ip}} \right) d\xi_{ia} d\xi_{ib}. \tag{2}$$

Then we may consider the hyperplane should be a Riemannian space. The structure of the hyperplane will be discussed by the several geometrical quantities. But we know that the induced metric on a unit-hypersphere in p-dimensional Euclidean is denoted as follows. Using a coordinate $(u_1, u_2, \ldots u_p)$ and

$$u_1 + u_2 + \cdots + u_p = 1, \quad u_a > 0 \ (a = 1, \ldots p),$$

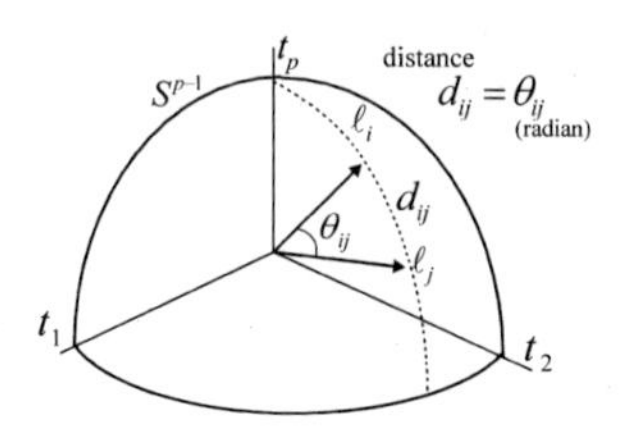

Fig. 1 Directional data

when we denote the hypersphere as follows:

$$\ell_1 = \sqrt{u_1}, \quad \ell_2 = \sqrt{u_2}, \quad \ldots, \quad \ell_{(p-1)} = \sqrt{u_{(p-1)}}, \quad \ell_p = \sqrt{1 - \sum_{b=1}^{p-1} u_b}$$

the induced metric is given by

$$ds^2 = \sum_{a=1}^{p-1}\sum_{b=1}^{p-1} g_{ab} du_a du_b = \sum_{a=1}^{p-1}\sum_{b=1}^{p-1} \frac{1}{4}\left(\delta_{ab}\frac{1}{u_a} + \frac{1}{u_p}\right) du_a du_b.$$

Then we know that the geometrical structure of the hyperplane (5) which has the line element (2) is a hypersphere.

From this result, we define a directional data, i.e. the data on the unit hypersphere using weighted $\boldsymbol{\xi}_i$ as

$$\ell_{ia} = \sqrt{\xi_{ia}}, \quad (a = 1, \ldots, p).$$

The main advantage using the data on the hypersphere is easy to get a global geodesic distance, because we know the geodesic curve on the hypersphere is the great circle (see Fig. 1). If we discuss on the hyperplane, we must get the geodesic curve, which is a solution of the geodesic equation.

3 Distribution on a Hypersphere

3.1 Descriptive Measures

Suppose the sample of n observations on a hypersphere $S^{(p-1)}$ is given by

$$\boldsymbol{\ell}_1, \boldsymbol{\ell}_2, \ldots, \boldsymbol{\ell}_n, \quad \boldsymbol{\ell}_i = (\ell_{i1}, \ell_{i2}, \ldots, \ell_{ip}), \quad \boldsymbol{\ell}_i'\boldsymbol{\ell}_i = 1.$$

Then, a mean direction is defined as

$$\bar{\boldsymbol{\ell}} = \sum_{i=1}^{n} \boldsymbol{\ell}_i / R,$$

where R is considered to be a total length of the data, that is defined as follows:

$$R^2 = \left(\sum_{i=1}^{n} \ell_{i1}\right)^2 + \left(\sum_{i=1}^{n} \ell_{i2}\right)^2 + \cdots + \left(\sum_{i=1}^{n} \ell_{ip}\right)^2.$$

The value of R becomes larger as n increase and also become larger when the data concentrate on the mean direction. On the other hand, R becomes smaller when the data is distributed according to uniform distribution. That is, R is regarded as the concentration measure. Consequently, a variance on the sphere can be defined as (Mardia 1972)

$$S = (n - R)/n, \quad 0 \leq R \leq n, \quad 0 < S < 1.$$

3.2 Distribution on a Hypersphere

Supposing $\boldsymbol{\ell} = (\ell_1, \ell_2, \ldots, \ell_p)$, $\boldsymbol{\ell}'\boldsymbol{\ell} = 1$ be a $(p-1)$-dimensional random variable on a hypersphere $S^{(p-1)}$, the density function is defined as a generalization of Fisher distribution by (Fisher 1959; Watson and Williams 1956)

$$f(\boldsymbol{\ell}; \boldsymbol{\lambda}, \kappa) = \frac{\kappa^{\frac{p}{2}-1}}{(2\pi)^{\frac{p}{2}} I_{\frac{p}{2}-1}(\kappa)} \exp[\kappa(\boldsymbol{\ell}'\boldsymbol{\lambda})], \ \kappa > 0, \tag{3}$$

where $\boldsymbol{\lambda}$ denotes the mean direction $\boldsymbol{\lambda} = (\lambda_1, \lambda_2, \ldots, \lambda_p)$, $\boldsymbol{\lambda}'\boldsymbol{\lambda} = 1$ and κ is considered to be a measure of concentration, i.e. variance parameter. $I_{\frac{p}{2}-1}(\kappa)$ is a modified Bessel function. This distribution should be a natural extension of multivariate normal distribution on a hypersphere. When we assume $p = 2$, the above distribution is known as von Mises distribution (von Mises 1981). The shape of the density function is shown in Fig. 2. In the distribution (3), $\boldsymbol{\lambda}$ and κ are the unknown parameters. If these parameters $\boldsymbol{\lambda}$ and κ are given, the distribution is determined uniquely. So we denote it $S(\boldsymbol{\lambda}, \ \kappa)$.

Suppose $\boldsymbol{\ell}_1, \ldots, \boldsymbol{\ell}_n$ are a set of n independent samples, $\boldsymbol{\ell}_i$ being drawn from $S(\boldsymbol{\lambda}, \ \kappa)$. We shall estimate κ and $\boldsymbol{\lambda}$ by the method of maximum likelihood. The likelihood function is

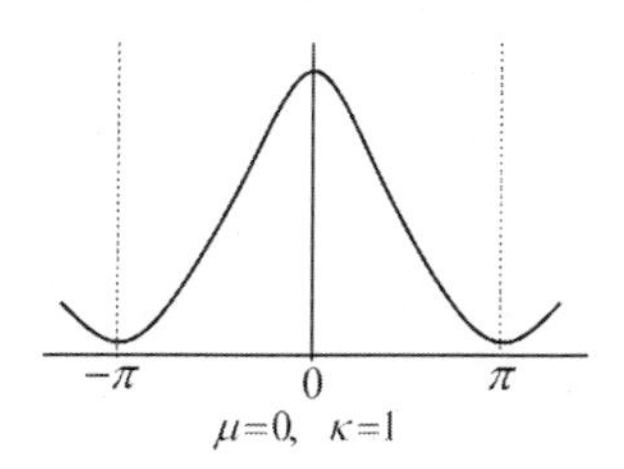

Fig. 2 von Mises distribution

$$\prod_{i=1}^{n} f(\boldsymbol{\ell}_i; \kappa, \boldsymbol{\lambda}) = \left[\frac{\kappa^{\frac{p}{2}-1}}{(2\pi)^{\frac{p}{2}} I_{\frac{p}{2}-1}(\kappa)}\right]^n \prod_{i=1}^{n} \exp[\kappa(\boldsymbol{\ell}_i'\boldsymbol{\lambda})].$$

Here, if we put

$$R_a = \sum_{i=1}^{n} \ell_{ia}, \quad R^2 = (R_1^2 + \cdots + R_p^2), \quad \bar{R} = \frac{R}{n},$$

the logarithm of the likelihood function is

$$\text{Const.} + n\left(\frac{p}{2} - 1\right)\log\kappa - n\log I_{\frac{p}{2}-1}(\kappa) + \kappa\sum_{a=1}^{p} \lambda_a R_a.$$

We may maximize this function with respect to $\boldsymbol{\lambda}$ and κ under the condition $\boldsymbol{\lambda}'\boldsymbol{\lambda} = 1$. Then we get

$$\hat{\lambda}_a = \frac{R_a}{R}, \quad a = 1, 2, \ldots p \quad \text{and} \quad \frac{I_{\frac{p}{2}}(\hat{\kappa})}{I_{\frac{p}{2}-1}(\hat{\kappa})} = \bar{R}$$

4 Discriminant Function

Suppose two populations Π_1 and Π_2 to be given on the hypersphere $S^{(p-1)}$. If we denote the density function of ath population is

$$f_a(\boldsymbol{\ell}; \boldsymbol{\lambda}_a, \kappa_a) = \frac{\kappa_a^{\frac{p}{2}-1}}{(2\pi)^{\frac{p}{2}} I_{\frac{p}{2}-1}(\kappa_a)} e^{\kappa_a(\boldsymbol{\ell}'\boldsymbol{\lambda}_a)} \equiv c(\kappa_a)\exp\{\kappa_a(\boldsymbol{\ell}'\boldsymbol{\lambda}_1)\}.$$

Then the Bayes decision rule is denoted as

$$\kappa_1(\boldsymbol{\ell}'\boldsymbol{\lambda}_1) + \log c(\kappa_1) + \log q_1 \geq \kappa_2(\boldsymbol{\ell}'\boldsymbol{\lambda}_2) + \log c(\kappa_2) + \log q_2 \quad \Rightarrow \boldsymbol{\ell} \in \Pi_1,$$
$$\kappa_1(\boldsymbol{\ell}'\boldsymbol{\lambda}_1) + \log c(\kappa_1) + \log q_1 < \kappa_2(\boldsymbol{\ell}'\boldsymbol{\lambda}_2) + \log c(\kappa_2) + \log q_2 \quad \Rightarrow \boldsymbol{\ell} \in \Pi_2.$$

When the parameter κ is common in two populations, i.e. $\kappa_1 = \kappa_2$, the rule is given by

$$\kappa(\boldsymbol{\ell}'\boldsymbol{\lambda}_1) + \log q_1 \geq \kappa(\boldsymbol{\ell}'\boldsymbol{\lambda}_2) + \log q_2 \quad \Rightarrow \ell \in \Pi_1,$$
$$\kappa(\boldsymbol{\ell}'\boldsymbol{\lambda}_1) + \log q_1 < \kappa(\boldsymbol{\ell}'\boldsymbol{\lambda}_2) + \log q_2 \quad \Rightarrow \boldsymbol{\ell} \in \Pi_2.$$

Furthermore, when the prior probabilities are equal, i.e. $q_1 = q_2$, the Bayes rule becomes as follows:

$$(\boldsymbol{\ell}'\boldsymbol{\lambda}_1) \geq (\boldsymbol{\ell}'\boldsymbol{\lambda}_2) \quad \Rightarrow \boldsymbol{\ell} \in \Pi_1,$$
$$(\boldsymbol{\ell}'\boldsymbol{\lambda}_1) < (\boldsymbol{\ell}'\boldsymbol{\lambda}_2) \quad \Rightarrow \boldsymbol{\ell} \in \Pi_2.$$

5 Numerical Experiments

The data used in this numerical experiment is "a database of faces and non-faces" which is constructed and opened by MIT Center for biological and computational learning (http://www.ai.mit.edu/projects/cbcl) An object is the monochrome image of face or non-face. Each image is displayed by $19 \times 19 = 361$ pixels. Each pixel takes the value in the range [0,255]. Then we may consider that the number of attributes is 361. However, as an numerical example, we transform each attribute value into binary value, that is, each image is represented by 1 or 2 taking account of the correlation ratio among 361 attributes values. Here we shall consider the discrimination problem between face image and non-face image. The number of training data and test data are given in Table 2. In these data the objects whose attribute values are all 1, i.e. the image is black, are eliminated from the training and test data. When we assume the distribution of the hypersphere for two groups, face and non-face, the discrimination using common parameter κ has the best result. The misdiscrimination rate for training data is about 8%, and for the test data is about 25% (see Table 3).

A comparison of discriminant procedures for binary variables has been discussed by Asparoukhov and Krzanowski (2001). In this paper, 13 methods of discrimination have treated. As a result, they denoted that the traditional method, LDF, QDF, Kernel density estimation and k-Nearest method are not extremely bad, but the method based on neural networks (Multi-Layer NN, LVQ) seem to bring rather good results.

Table 2 The number of training and test data

	Face	Non-face
Training data	2,429	2,472
Test data	472	2,058

Table 3 Misdiscrimination rate in each method (%)

	Training data		Test data	
	Face	Non-face	Face	Non-face
Disc. func. on sphere*	9.2	7.4	49.1	18.9
SVM(radial)	0	0	66.1	36.3
SVM(sigmoid)	0.5	0.5	47.5	45.7
SVM(polynomial)	0	0	83.7	27.3
LVQ(4)	3.4	1.4	65.9	34.5
LVQ(10)	2.3	1.3	62.3	42.2
LVQ(40)	1.9	0.9	71.2	33.0
LDF	0.4	0.2	59.0	40.0

*Using the common κ

So in this paper, we have shown the results of Support vector machine (SVM) (Vapnik 1998) and learning vector quantization (LVQ) (Kohonen 1995) using the same data. Table 3 show the misdiscrimination rate (%), where the term in parentheses of SVM is the kernel function and the number in the parentheses of LVQ shows the number of codebook vectors.

From the results in Table 3, we know that each result shows its characteristics. In the discriminant function on the hypersphere, although the prediction for the test data is not extremely good, the result seems to be comparatively stable. In the support vector machine that map to a higher dimensional space by the kernel trick, is clearly overfitting in every kernel functions. Then in SVM, we must introduce an adequate stopping rule for the learning. In the method of learning vector quantization, the selection of the initial codebook vectors and the number of them should be selected suitably, but it seems to be difficult to determine previously. The lase line in Table 3 shows the result of LDF. The result is interesting. This seems to be rather better than the sophisticated methods.

In any case, these discrimination problem is closely connected with the concept of distance. Then it will be interesting to construct the algorithm of SVM and LVQ on the hypersphere.

6 Concluding Remarks

In this paper, we have proposed a method to analyse the binary data on the hypersphere using the transformation binary data into directional data. Here, the question why we have to discuss on the hypersphere will occur. With regard to distance function, we can define, for instance, Kullback–Leibler divergence on the $(p-1)$-dimensional hyperplane

$$x_1 + x_2 + \cdots + x_p = 1, \ x_a > 0.$$

It will be known that the Kullback–Leibler divergence between infinitesimal different points becomes χ-square distance. We have shown that This is regarded as a Riemannian metric in $(p-1)$-dimensional space. And also, if we calculate the curvature of this space, we know it has a positive constant curvature. Then we may consider this space is sphere. Furthermore, the merit to discuss on the sphere, we know the geodesic curve is the great circle. Then we can get the global distance between any two points.

Recently, in the field of datamining, the nonparametric methods are highly used without need for any assumption of distribution. However, it will be better to use a parametric method, if possible. Because, it will be trivial that the performance of the prediction will be more decrease than parametric case, without any assumptions. This is also shown in the Table 3.

References

ASPAROUKHOV, O.K. and KRZANOWSKI, W.J. (2001): A comparison of discriminant procedures for binary variables. *Computational Statistics and Data Analysis* **38**, 139–160.

FISHER, R.A. (1959): *Statistical Methods and Scientific Inference,* 2nd edn. Oliver and Boyd. Edinburgh.

KOHONEN, T. (1995): Lerning Vector Quantization. In M.A. Arbib (ed.): *The Handbook of Brain Theory and Neural Networks.* MIT, Massachusetts, 537–540.

MARDIA, K. (1972): *Statistics of Directional Data*, Academic, New York.

VAPNIK, V.N. (1998): *Statistical Learning Theory.* Wiley, New York.

VOM MISES, R. (1981): Über die "Ganzzahligkeit" der Atomgewicht und verwandte Fragen. *Physikal. Z.*, **19**, 490–500.

WATSON, G.S. and WILLIAMS, E.J.(1956): On the construction of significance tests on the circle and shpere. *Biometrika* **43**, 344–352.

Fuzzy Clustering Based Regression with Attribute Weights

M. Sato-Ilic

Abstract Fuzzy regression methods are proposed considering classification structure which is obtained as a result of fuzzy clustering with respect to each attribute. The fuzzy clustering is based on dissimilarities over objects in the subspace of the object's space. Exploiting the degree of belongingness of objects to clusters with respect to attributes, we define two fuzzy regression methods in order to estimate the fuzzy cluster loadings and weighted regression coefficients. Numerical examples show the applicability of our proposed method.

1 Introduction

Regression analysis is well-known and widely used in many areas. In classical regression analysis, estimate of least squares is usually obtained under the assumption of homoscedasticity of variance for residuals. In order to solve the problem of heteroscedastic residuals, many weighted regression analyses have been proposed (Draper and Smith 1966; Dobson 1990). Involving weights for the regression of geographical data, a geographically weighted regression model (Brunsdon et al. 1998) has been proposed. Using the idea that weights are obtained by using the fuzzy degree of fuzzy clustering, a switching regression model has been proposed (Hathaway and Bezdek 1993). Also, we have proposed fuzzy cluster loading models in order to obtain interpretation of fuzzy clustering result and a fuzzy weighted regression model (Sato-Ilic 2003; Sato-Ilic and Jain 2006).

Due to the recent challenge to analyze data which has a larger number of attributes than the number of objects, a clustering method including the idea of

M. Sato-Ilic
Department of Risk Engineering, Faculty of Systems and Information Engineering,
University of Tsukuba, Tennodai 1-1-1, Tsukuba, Ibaraki 305-8573, Japan,
E-mail: mika@sk.tsukuba.ac.jp

A. Okada et al. (eds.), *Cooperation in Classification and Data Analysis*,
Studies in Classification, Data Analysis, and Knowledge Organizations.
DOI: 10.1007/978-3-642-00668-5_7,

subspaces of data have been discussed (Friedman and Meulman 2004). This method uses the weights of attributes with respect to clusters as well as other clustering methods (Frigui and Caudill 2006; Sato-Ilic 2003).

In order to consider the difference of attributes of objects to the fuzzy cluster loading model and a fuzzy weighted regression model, this paper propose two fuzzy regression methods involving weights of attributes. Human cognize classification using perceptional measure but this measure dose not always treat the all attributes equally. For example, if we need to classify several flowers with attributes of colors and shapes perceptively, then one may classify the flowers largely based on colors and the other may classify the flowers largely based on shapes. In this case, we naturally give the weights for showing the significance of each attribute.

In Sect. 2, we explain attribute based dissimilarity and show two attribute based fuzzy clustering methods (Sato-Ilic 2006). Sections 3 and 4 explain a fuzzy cluster loading model and a fuzzy weighted regression model. We propose a fuzzy cluster loading model considering weights of attributes and a fuzzy weighted regression model considering weights of attributes in Sect. 5. Section 6 shows a numerical example. We describe the conclusion in Sect. 7.

2 Fuzzy Clustering Considering Attributes

Suppose a set of n objects with respect to p attributes is as follows:

$$X = \{\boldsymbol{x}_1, \boldsymbol{x}_2, \ldots, \boldsymbol{x}_n\}, \quad \boldsymbol{x}_i = (x_{i1}, \ldots, x_{ip}), \ i = 1, \ldots, n.$$

We define a dissimilarity matrix D_a and dissimilarity between objects i and j with respect to an attribute a as follows:

$$D_a = (d_{ija}), \quad d_{ija} = (x_{ia} - x_{ja})^2, \quad i, j = 1, \ldots, n, \ a = 1, \ldots, p. \tag{1}$$

Then a dissimilarity $\tilde{d}_{ij}$ between objects i and j can be defined as follows:

$$\tilde{d}_{ij} = \sum_{a=1}^{p} w_a d_{ija}, \ i, j = 1, \ldots, n. \tag{2}$$

w_a shows a weight of an attribute a. In particular, conventional squared Euclidean distance $d_{ij} = \sum_{a=1}^{p}(x_{ia} - x_{ja})^2, \ i, j = 1, \ldots, n$ is the case when $w_a = 1, \ \forall a$ in (2). This means that squared Euclidean distance is a special case of $\tilde{d}_{ij}$ when we assume the same weight for all of the attributes in (2).

We define the objective function with respect to an attribute a by using (1) and FANNY algorithm (Kaufman and Rousseeuw 1990) as follows:

$$J(U_a) = \sum_{k=1}^{K} \left[\frac{\sum_{i=1}^{n} \sum_{j=1}^{n} (u_{ika})^m (u_{jka})^m d_{ija}}{2 \sum_{s=1}^{n} (u_{ska})^m} \right], \quad a = 1, \ldots, p, \tag{3}$$

where, u_{ika} shows degree of belongingness of an object i to a cluster k with respect to an attribute a and satisfies the following conditions:

$$u_{ika} \in [0, 1], \quad \sum_{k=1}^{K} u_{ika} = 1. \tag{4}$$

m, $(1 < m < \infty)$ shows a control parameter which can control fuzziness of the belongingness and in this algorithm, it is defined as $m = 2$. K is the number of clusters. u_{ika} can be estimated by minimizing (3).

For the second method, we define the following objective function

$$F(U_a, v) = \sum_{i=1}^{n} \sum_{k=1}^{K} \sum_{a=1}^{p} u_{ika}^m (x_{ia} - v_{ka})^2, \tag{5}$$

where, v_{ka} shows a center of a cluster k with respect to an attribute a. If $\exists i, k, u_{ika}$ is a constant for all a, then this is the same objective function as fuzzy c-means (Bezdek 1987). The purpose is to estimate $U_a = (u_{ika})$ and $v = (v_{ka})$ which minimize (5). From conditions shown in (4), the local extrema of (5) can be obtained as follows:

$$u_{ika} = 1 / \sum_{t=1}^{K} \left[\frac{x_{ia} - v_{ka}}{x_{ia} - v_{ta}} \right]^{\frac{2}{m-1}}, \quad v_{ka} = \frac{\sum_{i=1}^{n} u_{ika}^m x_{ia}}{\sum_{i=1}^{n} u_{ika}^m}. \tag{6}$$

We can estimate the degree of belongingness by alternating optimization using (6).

3 Fuzzy Cluster Loading Model

In order to obtain an interpretation of the fuzzy clustering result, we have proposed the following model:

$$\boldsymbol{u}_k = X \boldsymbol{z}_k + \boldsymbol{\varepsilon}_k, \quad k = 1, \ldots, K, \tag{7}$$

using

$$\boldsymbol{u}_k = \begin{pmatrix} u_{1k} \\ \vdots \\ u_{nk} \end{pmatrix}, \ X = \begin{pmatrix} x_{11} & \cdots & x_{1p} \\ \vdots & \ddots & \vdots \\ x_{n1} & \cdots & x_{np} \end{pmatrix}, \ \boldsymbol{z}_k = \begin{pmatrix} z_{1k} \\ \vdots \\ z_{pk} \end{pmatrix}, \ \boldsymbol{\varepsilon}_k = \begin{pmatrix} \varepsilon_{1k} \\ \vdots \\ \varepsilon_{nk} \end{pmatrix},$$

$$\varepsilon_{ik} \sim N(0, \sigma^2),$$

where, u_{ik} is degree of belongingness of an object i to a cluster k which is obtained by using a fuzzy clustering method and is assumed to satisfy the following conditions in order to avoid $u_{ik} = 0$:

$$u_{ik} \in (0, 1), \ \sum_{k=1}^{K} u_{ik} = 1. \tag{8}$$

ε_{ik} is an error under an assumption of $\varepsilon_{ik} \sim N(0, \sigma^2)$. z_{ak} shows the fuzzy degree which represents the amount of loading of a cluster k to an attribute a which we call fuzzy cluster loading. This parameter will show how each cluster can be explained by each variable.

The estimate of least squares of $\boldsymbol{z}_k$ for (7) is obtained as follows:

$$\tilde{\boldsymbol{z}}_k = (X^t X)^{-1} X^t \boldsymbol{u}_k, \tag{9}$$

by minimizing $\boldsymbol{\varepsilon}_k^t \boldsymbol{\varepsilon}_k = \varepsilon_{1k}^2 + \cdots + \varepsilon_{nk}^2$. Using

$$U_k = \begin{pmatrix} u_{1k}^{-1} & \cdots & 0 \\ \vdots & \ddots & \vdots \\ 0 & \cdots & u_{nk}^{-1} \end{pmatrix},$$

the model (7) can be rewritten again as

$$\mathbf{1} = U_k X \boldsymbol{z}_k + \boldsymbol{e}_k, \ \boldsymbol{e}_k \equiv U_k \boldsymbol{\varepsilon}_k, \ k = 1, \ldots, K, \tag{10}$$

where,

$$\mathbf{1} = \begin{pmatrix} 1 \\ \vdots \\ 1 \end{pmatrix}, \ \boldsymbol{e}_k = \begin{pmatrix} e_{1k} \\ \vdots \\ e_{nk} \end{pmatrix}, \ e_{ik} \sim N(0, \sigma^2).$$

By minimizing

$$\boldsymbol{e}_k^t \boldsymbol{e}_k = \boldsymbol{\varepsilon}_k^t U_k^2 \boldsymbol{\varepsilon}_k, \tag{11}$$

we obtain the estimate of least squares of $\boldsymbol{z}_k$ for model (10) as follows:

$$\tilde{\tilde{\boldsymbol{z}}}_k = (X^t U_k^2 X)^{-1} X^t U_k \mathbf{1}. \tag{12}$$

From (11), we can see that $\tilde{\tilde{z}}_k$ is the estimate of weighted least squares of z_k. Equation (11) can be rewritten as follows:

$$e_k^t e_k = \varepsilon_k^t U_k^2 \varepsilon_k = (u_{1k}^{-1}\varepsilon_{1k})^2 + \cdots + (u_{nk}^{-1}\varepsilon_{nk})^2. \tag{13}$$

In (13), from condition (8), a larger u_{ik} has a smaller $(u_{ik}^{-1}\varepsilon_{ik})^2$, since

$$a > b \Rightarrow (a^{-1}c)^2 < (b^{-1}c)^2, \ a, b, c > 0.$$

This means that if an object i belongs to the cluster k with a large degree, that is classification structure of the object i is close to crisp, then error related with the object i becomes smaller. So, $\tilde{\tilde{z}}_k$ is obtained considering not only the fitness of the model shown in (7) but also considering the classification structure.

4 A Weighted Regression Analysis Using Fuzzy Clustering

Using a fuzzy clustering result obtained as classification of the data consisting of independent variables, a model is defined as

$$\boldsymbol{y} = U_k^{-1} X \boldsymbol{\beta}_k + \tilde{\boldsymbol{e}}_k, \tag{14}$$

where

$$U_k^{-1} = \begin{pmatrix} u_{1k} & \cdots & 0 \\ \vdots & \ddots & \vdots \\ 0 & \cdots & u_{nk} \end{pmatrix}, \quad \boldsymbol{\beta}_k = \begin{pmatrix} \beta_{1k} \\ \vdots \\ \beta_{pk} \end{pmatrix}, \quad \tilde{\boldsymbol{e}}_k = \begin{pmatrix} \tilde{e}_{1k} \\ \vdots \\ \tilde{e}_{nk} \end{pmatrix}, \quad \tilde{e}_{ik} \sim N(0, \sigma^2).$$

$\boldsymbol{y}$ shows values of a dependent variable and X shows values of independent variables. The estimate of least squares of $\boldsymbol{\beta}_k$ in (14) is obtained as

$$\tilde{\boldsymbol{\beta}}_k = (X^t (U_k^{-1})^2 X)^{-1} X^t U_k^{-1} \boldsymbol{y}. \tag{15}$$

5 Attribute Based Fuzzy Cluster Loading Model and a Fuzzy Weighted Regression Analysis

Using the idea of fuzzy clustering result with respect to attributes shown in (4), we redefine the models shown in (10) and (14) as follows:

$$\mathbf{1} = \hat{X}_k \hat{\boldsymbol{z}}_k + \hat{\boldsymbol{e}}_k, \quad k = 1, \ldots, K, \tag{16}$$

where,

$$\hat{X}_k = \begin{pmatrix} u_{1k1}^{-1}x_{11} & \cdots & u_{1kp}^{-1}x_{1p} \\ \vdots & \ddots & \vdots \\ u_{nk1}^{-1}x_{n1} & \cdots & u_{nkp}^{-1}x_{np} \end{pmatrix}, \ \hat{\boldsymbol{z}}_k = \begin{pmatrix} \hat{z}_{1k} \\ \vdots \\ \hat{z}_{pk} \end{pmatrix}, \ \hat{\boldsymbol{e}}_k = \begin{pmatrix} \hat{e}_{1k} \\ \vdots \\ \hat{e}_{nk} \end{pmatrix}, \ \hat{e}_{ik} \sim N(0, \sigma^2).$$

$$\boldsymbol{y} = \tilde{\tilde{X}}_k \tilde{\tilde{\boldsymbol{\beta}}}_k + \tilde{\tilde{\boldsymbol{e}}}_k, \tag{17}$$

where

$$\tilde{\tilde{X}}_k = \begin{pmatrix} u_{1k1}x_{11} & \cdots & u_{1kp}x_{1p} \\ \vdots & \ddots & \vdots \\ u_{nk1}x_{n1} & \cdots & u_{nkp}x_{np} \end{pmatrix}, \ \tilde{\tilde{\boldsymbol{\beta}}}_k = \begin{pmatrix} \tilde{\tilde{\beta}}_{1k} \\ \vdots \\ \tilde{\tilde{\beta}}_{pk} \end{pmatrix}, \ \tilde{\tilde{\boldsymbol{e}}}_k = \begin{pmatrix} \tilde{\tilde{e}}_{1k} \\ \vdots \\ \tilde{\tilde{e}}_{nk} \end{pmatrix}, \ \tilde{\tilde{e}}_{ik} \sim N(0, \sigma^2).$$

By minimizing $\hat{\boldsymbol{e}}_k^t \hat{\boldsymbol{e}}_k$ and $\tilde{\tilde{\boldsymbol{e}}}_k^t \tilde{\tilde{\boldsymbol{e}}}_k$, we obtain the estimates of least squares of $\hat{\boldsymbol{z}}_k$ and $\tilde{\tilde{\boldsymbol{\beta}}}_k$ in models (16) and (17) as follows:

$$\hat{\hat{\boldsymbol{z}}}_k = (\hat{X}_k^t \hat{X}_k)^{-1} \hat{X}_k^t \mathbf{1}. \tag{18}$$

$$\hat{\tilde{\tilde{\boldsymbol{\beta}}}}_k = (\tilde{\tilde{X}}_k^t \tilde{\tilde{X}}_k)^{-1} \tilde{\tilde{X}}_k^t \boldsymbol{y}. \tag{19}$$

6 Numerical Example

The data is iris data (Fisher 1936) which consists of 150 samples of iris flowers with respect to four variables, sepal length, sepal width, petal length, and petal width. The samples are observed from three kinds of iris flowers, iris setosa, iris versicolor, and iris virginica. The number of clusters is assumed to be 3.

Figure 1 shows the result of the FANNY algorithm. In this figure, three axes show the three clusters and the locations of dots show the values of degree of

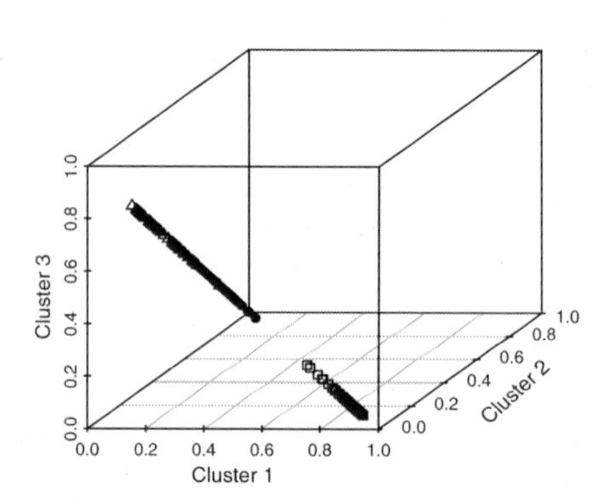

Fig. 1 Result of FANNY

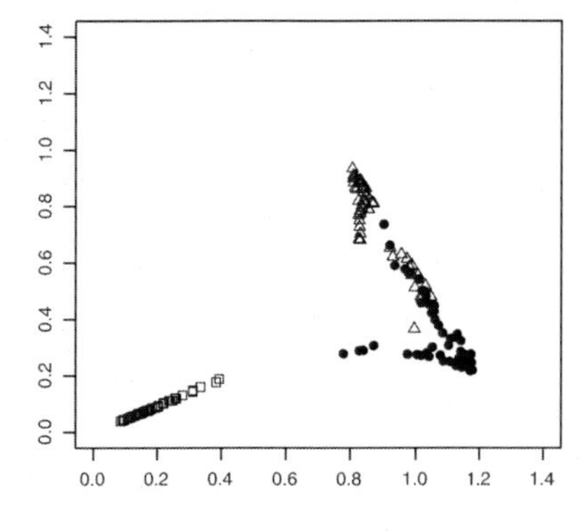

Fig. 2 Results of FANNY on solution plain

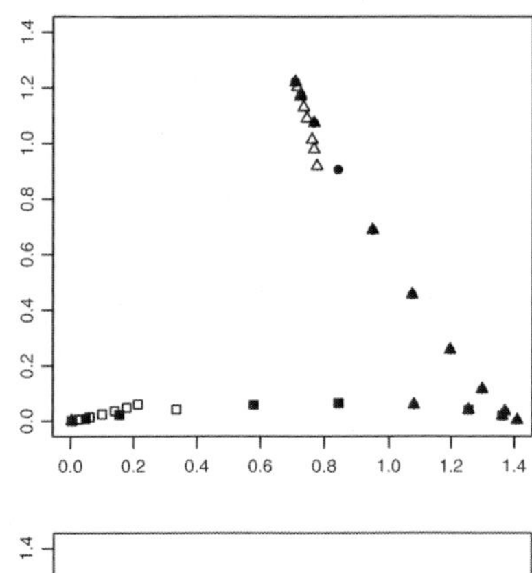

Fig. 3 Classification structure at sepal length

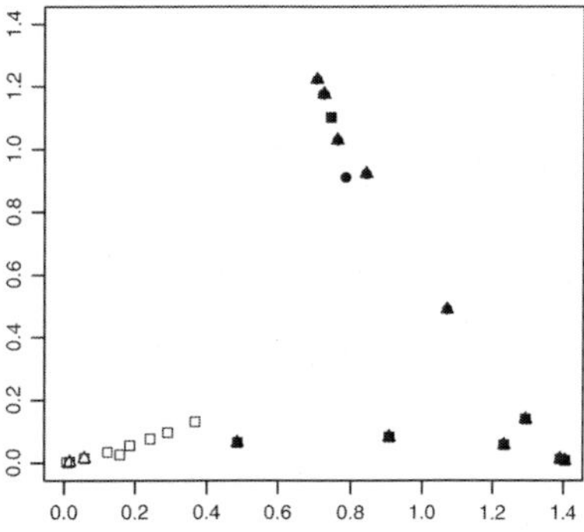

Fig. 4 Classification structure at sepal width

belongingness of each iris flower. In fact, due to the condition $\sum_{k=1}^{K} u_{ik} = 1$ shown in (8), the solutions u_{ik}, $(i = 1, \dots, 150,\ k = 1, \dots, 3)$ are only on the plain $\sum_{k=1}^{3} u_{ik} = 1$. Figure 2 shows situations of intersection between the space shown in Fig. 1 and the plain $\sum_{k=1}^{3} u_{ik} = 1$. Each symbol shows different kinds of iris flowers; squares show setosa, black dots show versicolor, and triangles show virginica. Figures 3–6 show the results of FANNY method when we use the dissimilarity matrixes D_1, D_2, D_3, and D_4 in (1). From these figures, we can see the difference of the classification structures with respect to each attribute.

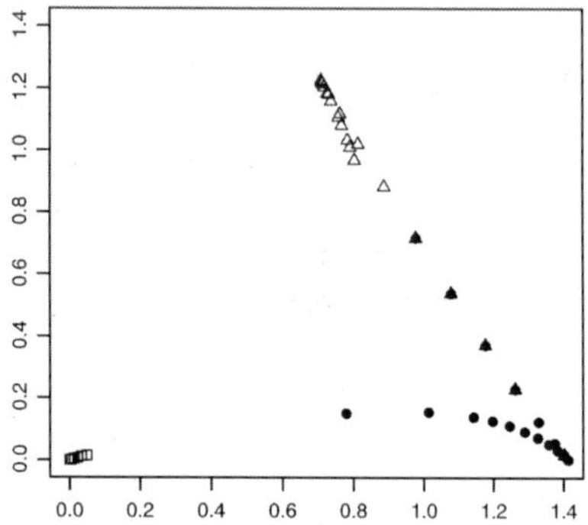

Fig. 5 Classification structure at petal length

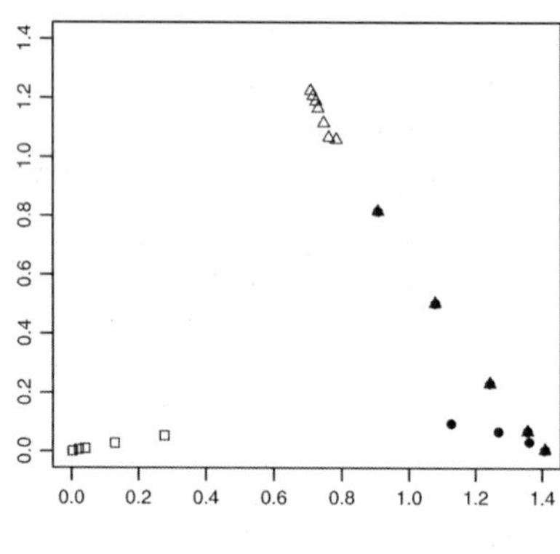

Fig. 6 Classification structure at petal width

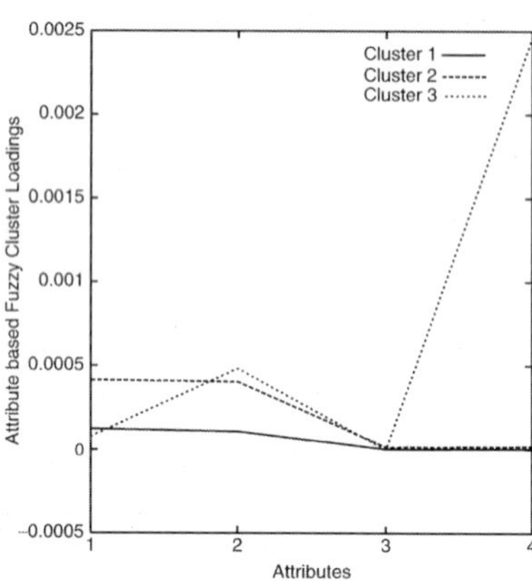

Fig. 7 Attribute based Fuzzy cluster loadings

Figure 7 shows the result of attribute based fuzzy cluster loadings shown in (18) and Fig. 8 shows the result of conventional fuzzy cluster loadings shown in (12). In these figures the abscissa shows attributes and ordinate shows values of fuzzy cluster loadings. From Fig. 7, we can see that cluster 2 is related with attributes of variables 1 and 2 which are sepal length and sepal width and is not related to petal length, and petal width. That is, cluster 2 is considered as a group created from an

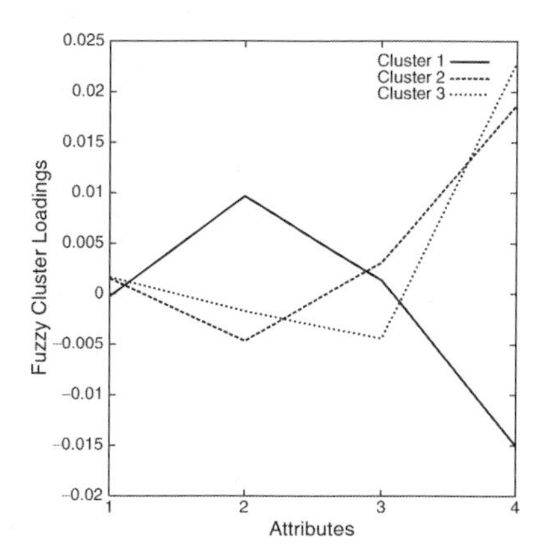

Fig. 8 Fuzzy cluster loadings

aspect of the sepal of flowers. Cluster 3 is related with variables 2 and 4 which show sepal width and petal width and is not related with sepal length and petal length. This means that cluster 3 is created from an aspect of attributes which are related with the width of flowers. From Fig. 8, we can not find any specific meanings related with the attributes.

7 Conclusion

We propose fuzzy regression methods considering attributes. These methods involve the consideration of classification structures in subspaces spanned by each attribute. Classification with the salience of attributes is adaptable for perceptional classification. Investigation of relationship with the cognitive psychology is a future problem.

References

BEZDEK, J.C. (1987): *Pattern Recognition with Fuzzy Objective Function Algorithms*. Plenum, New York.

BRUNSDON, C. FOTHERINGHAM, C. and CHARLTON, M. (1998): Geographically Weighted Regression Modeling Spatial Non-Stationarity. *Journal of the Royal Statistical Society*, 47, 431–443.

DOBSON, A.J. (1990): *An Introduction to Generalized Linear Models*. Chapman and Hall, London.

DRAPER, N.R. and SMITH, H. (1966): *Applied Regression Analysis*. Wiley, New York.

FISHER, R.A. (1936): The Use of Multiple Measurements in Taxonomic Problems. *Annals of Eugenics, 7,* 179–188.

FRIEDMAN, J.H. and MEULMAN, J.J. (2004): Clustering Objects on Subsets of Attributes. *Journal of the Royal Statistical Society, Series B, 66,* 815–849.

FRIGUI, H. and CAUDILL, J. (2006): Unsupervised Image Segmentation and Annotation for Content-Based Image Retrieval. *IEEE International Conference on Fuzzy Systems,* 274–279.

HATHAWAY, R.J. and BEZDEK, J.C. (1993): Switching Regression Models and Fuzzy Clustering. *IEEE Transactions Fuzzy Systems, 1, 3,* 195–204.

KAUFMAN, L. and ROUSSEEUW, P.J. (1990): *Finding Groups in Data.* Wiley, New York.

SATO-ILIC, M. (2003): On Kernel based Fuzzy Cluster Loadings with the Interpretation of the Fuzzy Clustering Result. *International Journal of Computational and Numerical Analysis and Applications, 4, 3,* 265–278.

SATO-ILIC, M. (2006): Fuzzy Clustering Considering Weights of Variables. *22nd Fuzzy System Symposium,* 167–168 (in Japanese).

SATO-ILIC, M. and JAIN, L.C. (2006): *Innovations in Fuzzy Clustering*. Springer, Berlin.

Polynomial Regression on a Dependent Variable with Immeasurable Observations

H.-Y. Siew and Y. Baba

Abstract In regression analysis, we sometimes encounter some detectable but immeasurable observations due to their extremely small values. These data are treated as binary data, which have value 1 or 0 for "exists" or "does not exist". The analysis can be carried out by using the methods of impulation, truncation and likelihood. In this paper, we study the fitting of these methods to a polynomial regression model. Furthermore, we examine the tendency of hypothesis test and information criteria in choosing the true model.

1 Introduction

While collecting data, we may encounter a situation where some observations are immeasurable due to the accuracy of measurement. These observations are sometimes treated as missing or censored. Regression analysis for data with missing values is a familiar topic to most statisticians. For instance, biologists may have difficulties in measuring the size of some microorganisms below a threshold, even under a microscope. Measuring equipment may be melted down while examining the effect of temperature to some devices. As a further example, Tobin (1958) proposed the Tobit model to analyze the expenditure data of households, where the dependent variable, expenditure, has a lower limiting value, usually zero. In such cases, it is known that the unobserved points are not outliers due to a mistake like a typing error and there are actual underlying values for these observations. Therefore, we should keep them in the analysis. Some literature references related to this topic include Baba (2003), Little and Rubin (2002) and Rubin (1976a, 1976b).

In this paper, we focus on a polynomial regression which consists of some immeasurable observations due to theirextremely small values. Since an observa-

H.-Y. Siew(✉)
The Institute of Statistical Mathematics, 4-6-7 Minami Azabu, Minato-ku, Tokyo 106-8569, Japan, E-mail: haiyen79@gmail.com

A. Okada et al. (eds.), *Cooperation in Classification and Data Analysis*,
Studies in Classification, Data Analysis, and Knowledge Organizations.
DOI: 10.1007/978-3-642-00668-5_8,

tion may or may not exist in the immeasurable region, we define the observations falling below a threshold, U, as

$$U = \begin{cases} 1, & \text{exist,} \\ 0, & \text{not exist.} \end{cases}$$

For observations lying above the threshold, we can measure the responses numerically. Therefore, we have both binary and continuous data for observations lying below and above the threshold in our analysis. In Sect. 2, we transform the response variable by taking logarithm and define a polynomial regression model for the transformed variable. Then, a mixed likelihood function is defined in Sect. 3. Section 4 gives two simulated examples based on 100 simulated data sets. The responses are generated based on a quadratic model with random errors. The random error variable is assumed to be normally distributed with mean 0 and variance σ^2. The fitting of the mixed likelihood model is compared to the imputation and truncated models. The imputation model is developed based on the imputation method by assigning a value of half of the threshold to each immeasurable point. For the truncated model, the binary data is discarded and only the continuous data is analyzed. Moreover, the tendency of hypothesis test and information criteria in choosing a true model when σ varies will be shown, too. Section 5 contains some concluding remarks.

2 Preliminaries

Let Z be a random variable $\in$, which is immeasurable when taking values below a threshold, $Z = z_0$. Thus, Z can be regarded as

$$Z = \begin{cases} z_0/2, & 0 < z < z_0, \\ z, & z \geq z_0. \end{cases} \tag{1}$$

Since Z may take values which are too small to be measured, we transform it into Y, where $Y = \log Z$.

Now, let us consider a polynomial regression model,

$$\boldsymbol{y} = \boldsymbol{x}\boldsymbol{\beta} + \boldsymbol{\varepsilon},$$

where

$$\boldsymbol{y} = (y_1, \ldots, y_n)',$$

$$\boldsymbol{x} = \begin{pmatrix} 1 & x_1 & \cdots & x_1^k \\ 1 & x_2 & \cdots & x_2^k \\ \vdots & & & \\ 1 & x_n & \cdots & x_n^k \end{pmatrix},$$

$$\boldsymbol{\beta} = (\beta_0, \beta_1, \ldots, \beta_k)',$$

where $\boldsymbol{Y}$ is a data vector of the dependent variable transformed from Z, $\boldsymbol{x}$ is a data matrix of independent variables $\boldsymbol{X}, \boldsymbol{X}^2, \ldots, \boldsymbol{X}^2$, and $\boldsymbol{\beta}$ is a vector of unknown coefficients. We also assume that $\boldsymbol{\varepsilon} = (\varepsilon_1, \ldots, \varepsilon_n)'$ is the random error vector having the normal distribution $N(\boldsymbol{0}, \sigma^2 \boldsymbol{I}_n)$, where $\boldsymbol{I}_n$ is an $n \times n$ identity matrix.

3 Likelihood Functions

Suppose that we have sample of size n, which has m_1 and m_2 binary and continuous data for the dependent variable, $\boldsymbol{Y}$. Then the data matrix of the independent variables, $\boldsymbol{x}$, and the data vector of the dependent variable, $\boldsymbol{y}$, are partitioned into two groups as following:

$$\boldsymbol{x} = (\boldsymbol{x}_1, \boldsymbol{x}_2)'$$

and

$$\boldsymbol{y} = (\boldsymbol{y}_1, \boldsymbol{y}_2)',$$

where $(\boldsymbol{x}_1, \boldsymbol{y}_1)$ and $(\boldsymbol{x}_2, \boldsymbol{y}_2)$ are the corresponding vectors of the observations lying below and above a threshold, respectively.

Let $f(t)$ and $\Phi(t)$ be the probability density and distribution functions of a standard normal distribution. Then given $\boldsymbol{x}_j$, the probability functions of an observation lying below and above the threshold are

$$\Pr\left(y < y_0 | \boldsymbol{x}_j\right) = \Phi\left(\frac{y_0 - \boldsymbol{x}_j \boldsymbol{\beta}}{\sigma}\right),$$

and

$$\Pr\left(y > y_0 | \boldsymbol{x}_j\right) = 1 - \Phi\left(\frac{y_0 - \boldsymbol{x}_j \boldsymbol{\beta}}{\sigma}\right),$$

where $y_0 = \log z_0$ is the threshold of $\boldsymbol{Y}$. Then, a simple likelihood function is defined as

$$L_{\text{simple}}(\boldsymbol{\beta}, \sigma | \boldsymbol{x}) = \prod_{j=1}^{m_1} \Phi\left(\frac{y_0 - \boldsymbol{x}_{1j}\boldsymbol{\beta}}{\sigma}\right) \prod_{j'=1}^{m_2} \left[1 - \Phi\left(\frac{y_0 - \boldsymbol{x}_{2j'}\boldsymbol{\beta}}{\sigma}\right)\right].$$

For the observations lying above the threshold, the response $\boldsymbol{y}_2$ is actually normally distributed with a mean $\boldsymbol{x}_2\boldsymbol{\beta}$ and variance-covariance matrix $\sigma^2 \boldsymbol{I}_{m_2}$. Therefore, we define a mixed likelihood function as

$$L_{\text{mixed}}(\boldsymbol{\beta}, \sigma | \boldsymbol{x}) = \prod_{j=1}^{m_1} \Phi\left(\frac{y_0 - \boldsymbol{x}_{1j}\boldsymbol{\beta}}{\sigma}\right) \prod_{j'=1}^{m_2} f\left(\frac{y_{j'} - \boldsymbol{x}_{2j'}\boldsymbol{\beta}}{\sigma}\right), \tag{2}$$

where $\boldsymbol{x}_{ij}$ is the jth row of the data matrix $\boldsymbol{x}_i$, for $i = 1, 2$. Note that the simple likelihood function is inferior to the mixed likelihood function because it concerns only the probabilities of the observations lying above or below the thresfold. On the contrary, the mixed likelihood function takes into account not only the existence of the binary data, but also the observed values of the continuous data. Therefore, the analysis should yield better results.

Since the mixed likelihood function involves the distribution function of the standard normal distribution, there is no closed form solution for the maximum likelihood estimators. Therefore, we use an approximation to the distribution function

$$\Phi(x) \approx \frac{e^{2y}}{1+e^{2y}},$$

where $y = a_1 x\left(1 + a_2 x^2\right), a_1 = 0.7988$ and $a_2 = 0.04417$, which is proposed by Page (1977), with the maximum error of 0.14×10^{-3}.

4 Simulation Studies

In this example, we assume that σ is known. For 100 simulated samples of size $n = 31$, which are based on a model $y = 4 - x + 0.05x^2 + \varepsilon$, where $\varepsilon \sim N(0, 1)$, the threshold is set at $y = 0$. On average, there are about 8 observations lying below the threshold, which are assumed to be immeasurable points. For each sample, we compare the performance of the quadratic and cubic fits for the imputation, truncated and mixed likelihood models. From (1), the dependent variable of the imputation model is given as

$$\boldsymbol{Y} = \begin{cases} -\log 2, & y < 0, \\ y, & y >= 0. \end{cases}$$

The results are shown in Fig. 1 and Table 1. Figure 1 contains the plots of quadratic and cubic fits for 100 samples for imputation, truncated and mixed likelihood models, respectively. The black dashed curves in each plot indicate the true model. From the figures, it is clear that the fits for imputation model are comparatively poor and the fits for the truncated model look over-estimated. In overall, both quadratic and cubic fits for the mixed likelihood model fluctuate more closely about the true model, compared with other models. Table 1 gives the average of the estimates of the parameter vectors for the 100 samples. The numerical outcomes are consistent with the figures. The estimates of the mixed likelihood model for both quadratic and cubic fits are most accurate and unbiased among all.

Next, for each dataset σ is estimated by the square root of the mean square error of the continuous data, which is given by

$$\text{MSE} = \frac{\boldsymbol{Y}_2'\boldsymbol{Y}_2 - \widehat{\boldsymbol{\beta}}'\boldsymbol{X}_2'\boldsymbol{Y}_2}{m_2 - p}.$$

The means and variances of the estimates of the σ for each model are shown in Table 2. The estimator of the mixed likelihood model seems to be unbiased with relatively small value of variance. The imputation model has the minimum variance because the same value is assigned to each immeasurable observation.

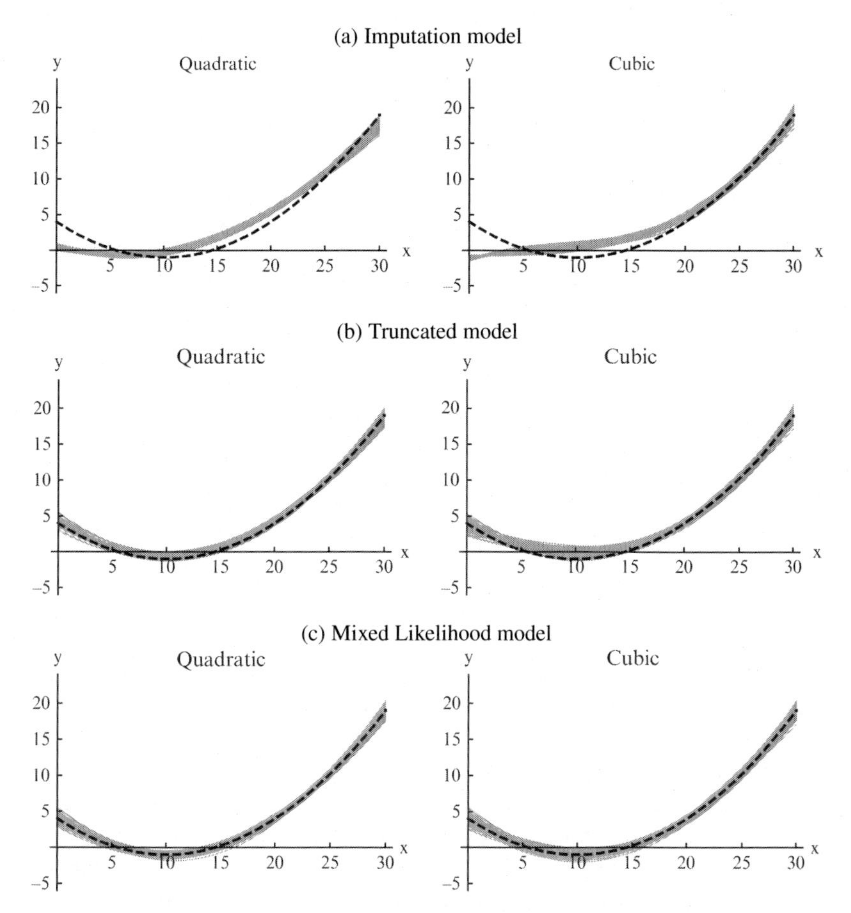

Fig. 1 Plots of 100 quadratic and cubic fits for imputation, truncated and mixed likelihood models. The *black dashed curves* in each plot indicate the true model

Table 1 Means of the estimates of parameter vector $\boldsymbol{\beta}$ for each model. The parameters are estimated based on the quadratic- and cubic-type models

Model	Quadratic	Cubic
Imputation	(0.3379, −0.3766, 0.0317)	(−1.1264, 0.2620, −0.0224, 0.0012)
Truncated	(4.2404, −0.9588, 0.0480)	(3.7665, −0.7045, 0.0271, 0.0004)
Mixed likelihood	(4.0765, −1.0022, 0.0499)	(4.0393, −0.9844, 0.0485, 0)

Furthermore, we examined the significance of the cubic regressor for the mixed likelihood model for 100 samples by using a hypothesis test and the information criteria, AIC (Parzen et al. 1998) and AICc (Hurvich and Tsai 1989) to select the optimal models, between quadratic- and cubic-type models. The test statistic and

Table 2 Means and variances of the estimates of σ for each model

Model	Mean($\widehat{\sigma}$)		Variance($\widehat{\sigma}$)	
	Quadratic	Cubic	Quadratic	Cubic
Imputation	1.9650	2.0640	0.0480	0.0560
Truncated	0.9563	0.9217	0.0179	0.0173
Mixed likelihood	0.9991	0.9968	0.0213	0.0225

Table 3 Percentages of quadratic- and cubic-type mixed likelihood models being selected from 100 samples by t-test, AIC and AICc when $\sigma = 1$

Type of model	t-Test($\alpha = 5\%$)	AIC	AICc
Quadratic	0.91	0.84	0.94
Cubic	0.09	0.16	0.06

the criteria are given as

$$t = \frac{\widehat{\beta}_3}{se\left(\widehat{\beta}_3\right)}, \tag{3}$$

$$\text{AIC} = -2\text{MLL} + 2p,$$

$$\text{AICc} = \text{AIC} + \frac{2p\,(p+1)}{n-p-1},$$

where $\widehat{\beta}_3$ is the estimated cubic regression coefficient, se is the standard error, MLL is the maximum value of the log-likelihood function, n is the sample size and p is the number of parameters to be estimated. AICc is a criterion for small-sample bias adjustment. When $n \to \infty$, AICc converges to AIC. For the hypothesis test, cubic model is adequate if the absolute value of (3) is larger than the quantiles of $t_{(n-p)}$ distribution. On the other hand, a model with the minimum criterion is selected as the optimal model. Table 3 shows the percentage of the quadratic- and cubic-type models being selected by using the hypothesis test and the information criteria. All methods show that the quadratic model has a much higher probability of being selected as the best fitting model. This result is desirable since the data are generated by a quadratic model.

The accuracy of the information criteria we use in choosing the true model is determined by the value of σ. From (2), it is obvious that the increment in σ will reduce the value of maximum log-likelihood. This is because the fluctuations of the observations become more rapid when σ is large. Meanwhile, the percentage of the hypothesis test of accepting a wrong model depends on the significance level. Therefore, it is independent of σ. For illustration, 10,000 samples are simulated based on the similar model, for each σ varies from 0.5 to 5. The result is shown in Fig. 2, where the average percentages of accepting true model by the three methods with error bars are drawn. At 5% α-level, the two-sided t-test has about 97% of accepting true model for all σ. On the other hand, the frequency of selecting quadratic model

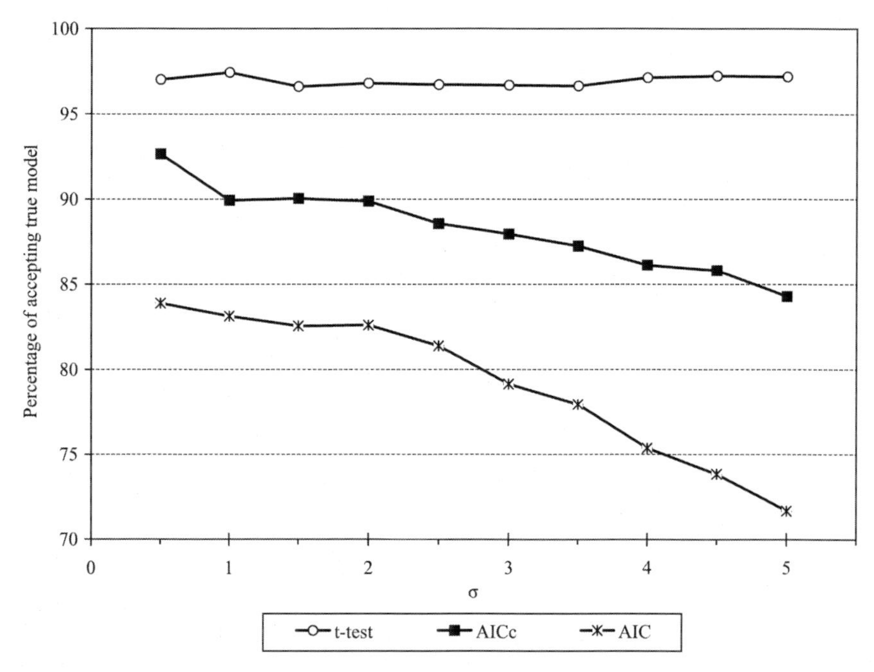

Fig. 2 Percentages of t-test, AIC and AICc accepting true model with error bars of one standard deviation using mixed likelihood models for various σ

using the information criteria decreases as σ increases, which is in agreement with the stated result. From these two examples, we notice that AIC always has lower percentage in selecting quadratic model due to its tendency for choosing a higher level model.

5 Conclusion

From the examples, we conclude that the mixed likelihood model is the optimal model among all. After examining the significance of a particular regression coefficient, the chosen model can be used to estimate the immeasurable points. In addition, the presented techniques are applicable to the case of large extreme values and multivariate regression problems.

Acknowledgements We are grateful to the referees and editors for careful reading and insightful comments

References

AKAIKE, H. (1974): A New Lookat the Statistical Model Identification. IEEE Transactions on Automatic Control *19,* 716–723.

BABA, Y. (2003): Regression Analysis of Data with Observation Limit. In *Bulletin of the International Statistical Institute, 54th Session Contributed Papers Book, 1,* 55–56.

HURVICH, C.M. and TSAI, C.L. (1989): Regression and Time Series Model Selection in Small Samples. *Biometrika, 76,* 297–307.

LITTLE, R.J.A. and RUBIN, D.B. (2002): *Statistical Analysis with Missing Data.* Wiley, New Jersey.

PAGE, E. (1977): Approximation to the Cumulative Normal Function and Its Inverse for Use on a Pocket Calculator. *Applied Statistics, 26,* 75–76.

RUBIN, D.B. (1976a): Comparing Regressions When Some Predictor Values Are Missing. *Technometrics, 18,* 201–205.

RUBIN, D.B. (1976b): Inference and Missing Data. *Biometrika, 63,* 581–592.

TOBIN, J. (1958): Estimation of Relationships for Limited Dependent Varibles. Econometrica, *26,* 24–36.

Part II
Methods in Fields

Feedback Options for a Personal News Recommendation Tool

C. Bomhardt and W. Gaul

Abstract Recommendations can help to vanquish the information overload problem on the web. Several websites provide recommendations for their visitors. If users desire recommendations for sites without this service, they can use browsing agents that give recommendations. In order to obtain user profiles, common agents store interesting and/or uninteresting web pages, use them as training data for the construction of classifiers, and give recommendations for unseen web pages. Feedback via explicit rating is regarded as most reliable but exhausting method to obtain training data. We present three alternative feedback options (explicit, implicit, and hybrid) and evaluate the alternatives via SVMs. We show that feedback options that are less exhausting than explicit rating can be applied successfully.

1 Introduction

Information overload is a severe problem that influences web usage. It arises out of the sheer mass of available information on the web and can even worsen due to technical reasons. Welcome pages of several online news websites present lists of headlines and abstracts of articles. A user has to click on, e.g., a headline and wait until the web page containing the accompanying article is transferred and displayed. In order to read another article, users often have to navigate back to the welcome page. Compared to fast running over the pages of a classical newspaper, information gathering via online news websites can be time consuming, less intuitive, and uncomfortable. This problem is not limited to online newspapers but affects also other websites like online shops or portals. Several websites use personalization techniques to address this problem. According to Mobasher et al. (2000), web personalization

W. Gaul(✉)
Institut für Entscheidungstheorie und Unternehmensforschung, Universität Karlsruhe (TH), 76128 Karlsruhe, Germany, E-mail: wolfgang.gaul@wiwi.uni-karlsruhe.de

A. Okada et al. (eds.), *Cooperation in Classification and Data Analysis*, Studies in Classification, Data Analysis, and Knowledge Organizations.
DOI: 10.1007/978-3-642-00668-5_9,

can be described as any action that makes the web experience of a user personalized to the user's taste? One can further distinguish between personalization by information filtering and personalization by information supplementing. Most online shops offer information filtering via a product search function (a user can supply keywords and the shop presents a filtered product list) and automatic information supplementing (the shop recommends additional or alternative products). Some online newspapers allow their users to subscribe to news of predefined categories. All these examples have in common that they are offered by website operators and are limited to websites that provide this kind of service. If a user desires recommendations on a website without such a service, additional tools like browsing agents are necessary.

2 Interest Profiles

Interest profiles of some kind are required by every browsing agent. Some agents model user interests with the help of user-supplied keywords. Unseen web pages are recommended, if they are similar enough to the specified keywords (filtering-based approach). Other agents model user interests with the help of collections of interesting and/or uninteresting web pages. Classifiers are trained on these collections and used to obtain recommendations for unseen web pages (machine-learning based approach). Training pages are needed for the machine-learning based approach. Typical agents therefor allow their users to explicitly rate the page currently shown. This is regarded as most reliable method to obtain feedback but exhausting for the user. A more convenient way is to monitor user behavior and thereby collect interesting and uninteresting web pages.

Browsing agents do exist for various applications. A good overview is given in Middleton (2001). NewsWeeder (Lang 1995) is an agent for Usenet newsgroups. It is realized as web based newsgroup client, uses explicit feedback in order to learn preferences, and compiles personalized news collections. WebMate (Chen and Sycara 1998) is a personal browsing agent that tracks interesting documents via explicit feedback. A new document is recommended if it is similar enough to an interesting reference document. Personal WebWatcher (PWW) (Mladenić 2001) accompanies a single user to become a specialist concerning the interests of the corresponding person. PWW records the URLs of pages read by the user and considers web pages shown as interesting. WebWatcher (Joachims et al. 1997) collects interest profiles of several users based on keywords. It recommends web pages that were interesting for other people with similar interests and relies on implicit (link followed) and explicit (users leaving the website can tell if their tour was successful or not) feedback. The browsing agent NewsRec (Bomhardt 2004) is specialized in recommendations for online newspapers and uses explicit feedback.

3 Feedback Options

Today's machine-learning based browsing agents typically offer explicit feedback (NewsWeeder, WebMate, NewsRec) or track shown web pages (PWW) and consider them interesting (implicit feedback). This kind of implicit feedback leads to mislabeled documents if users request uninteresting web pages, a rather common situation. Other implicit interest indicators proposed by Claypool et al. (2001), Kelly and Belkin (2004), Kelly and Teevan (2003) include display time, scrolling activity, mouse activity, keyboard activity, bookmarking, and saving or printing of web pages. Thereof, display time is, according to literature, the most promising. While Claypool et al. (2001) found display time to be a good indicator, it turned out to be unsuitable in Kelly and Belkin (2004). We expect display time to be inadequate for personal news recommendations due to three reasons: (1) most articles are short, thus, the variance concerning transfer delays can outweigh the variance concerning viewing time of a web page, (2) in contrast to the basic assumption that long display times indicate interestingness, very short articles ("new security update available") can be interesting, and (3) telephone calls or other external interruptions can lead to artificially long display times during regular usage outside of a laboratory. Additional problems of some of the mentioned indicators concern their availability w.r.t. regular browsers (e.g., keyboard activity) or that most classifiers require positive and negative training examples, but not every indicator identifies interesting AND uninteresting content (e.g., articles in web pages not printed are not necessarily uninteresting).

Our goal was to reduce the burden of explicit feedback and increase the number of correctly labeled training documents as compared to implicit feedback. The main problem of implicit feedback, as implemented by PWW, is that every web page shown is considered as interesting, even if it is uninteresting. This problem is mitigated by our hybrid feedback approach.

Hybrid feedback combines implicit and explicit feedback in such a way that implicit feedback is superimposed by explicit feedback because users can explicitly rate pages seen as uninteresting.

Hybrid feedback reduces the burden of rating articles. Mislabeled training examples can still occur if unseen pages that are automatically considered as uninteresting are indeed interesting. However, this situation should not occur too often in a news recommendation application, as it can be assumed that users of a special interest news website set value on reading all interesting articles.

Feedback options require information about whether an article was requested (yes/no) and/or which explicit user rating was assigned (none, +, −) in order to incorporate it into a set of training documents. All possible combinations together with the true rating and the corresponding assignments to one of the training sets are contained in Table 1. Six cases are possible. It is assumed that a rating, if given, is valid and matches the true rating (cases 3 and 4 of Table 1). Cases 1 and 2 describe situations where articles were seen but not rated. According to the true rating, in case 1 the corresponding article is interesting, in case 2 it is not. The implicit and hybrid feedback options both assign the corresponding articles to the set of

Table 1 Assignment of training documents for different feedback options

Case	Article requested?	User rating	*True rating*	Training set based on feedback option		
				Explicit	Implicit	Hybrid
1	Yes	None	+	None	+	+
2	Yes	None	−	None	+	+
3	Yes	+	+	+	+	+
4	Yes	−	−	−	+	−
5	No	None	+	None	−	−
6	No	None	−	None	−	−

interesting documents, explicit feedback doesn't assign. In case 3 the article was seen, obtained a positive rating, and is interesting. All feedback options assign this article to the set of interesting training documents. In case 4, the article was seen, obtained a negative rating and is uninteresting. Explicit and hybrid feedback assign it to the set of uninteresting training documents whereas implicit feedback assigns it to the set of interesting training documents. In cases 5 and 6 the corresponding articles were not seen and, thus, couldn't obtain a user rating. In case 5 the article is interesting, in case 6 the article is uninteresting for the user. Explicit feedback doesn't assign the articles to any training set; hybrid and implicit feedback assign both articles to the set of uninteresting training documents. As one can see from Table 1, explicit feedback collects two training documents compared to the 6 documents of implicit and hybrid feedback. Explicit feedback has no mislabeled training documents, implicit feedback 3 (50% of the cases shown in Table 1) and hybrid feedback 2 (33%). For implicit feedback, no user rating action is required to obtain the set of training documents as all requested articles are labeled as interesting and the rest as uninteresting. For explicit feedback, only articles that are requested can be rated. Here, 2 rating actions occur. Within hybrid feedback only 1 rating action (a user who knows how hybrid feedback works doesn't have to rate in case 3) is necessary. If a user correctly handles a system with hybrid feedback, the situation of case 2 should not occur (because the user knows that without an user rating a requested but uninteresting article will be assigned to the set of interesting training documents by hybrid feedback). Thus, the error rate of hybrid feedback drops down to 16% of the cases shown in Table 1. This comparison shows that hybrid feedback requires less user ratings compared to explicit feedback and can lead to more training documents. Compared to implicit feedback, it misassigns less documents.

4 Web Page Classification

Due to space restrictions, we are limited to a brief overview of web page classification as used by NewsRec (see Bomhardt 2004 for further details). Typical web pages consist of HTML code containing the article text and formatting instructions, navigational elements, and advertisements (see Fig. 1). A web page should

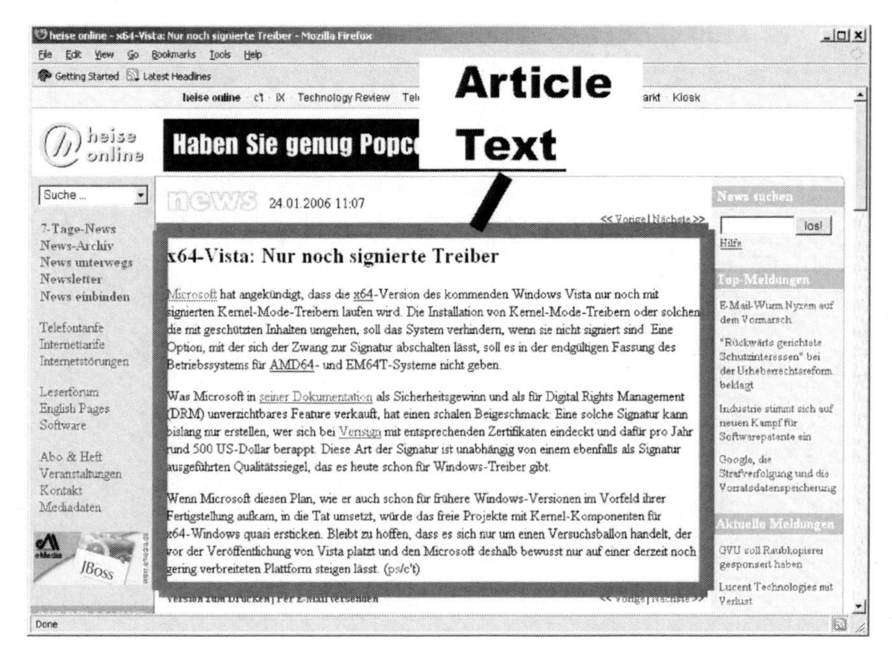

Fig. 1 A screenshot of the examined news website with highlighted article text

be classified based on its content without framing elements. In this context, the content is the article text. Its generic extraction is a problem on its own (Kushmerick et al. 1997). Thus, a wrapper was build manually. The extracted text still is contaminated with HTML tags like *<B>* that have to be removed. The remaining text is converted to lowercase and transformed to vector space model representation. As SVMs are used for classification, dimensionality reduction is not required. For every pre-processing step (*frequency transformation* [TF (1), LOG (2), BIN (3)], *term weighting* [NOWEIGHTS (4) or IDF (5)], and *normalization* [NONE (6), L2-NORM (7)]), one possibility has to be selected. We checked all possible combinations of these settings (see the first column of Table 2 for the application of different pre-processing settings). Further settings do exist but are not considered here as this work focuses on feedback options rather than pre-processing settings for text classification. Our pre-processing steps include settings recommended by Joachims (2002) for text classification with SVMs.

5 Evaluation Method

We selected recall (rec), precision (prec), and F1 as evaluation measures for web page classification. It should be mentioned that recall can be calculated here due to the known total number of news articles on a given website. This is not true

Table 2 Comparison of the feedback options. Column TWS (term weighting-scheme) contains the code of selected pre-processing settings explained in part 4 of this paper

TWS	Explicit			Implicit			Hybrid		
	Rec	Prec	F1	Rec	Prec	F1	Rec	Prec	F1
1-4-6	47.24%	60.99%	0.5324	70.50%	51.31%	0.5939	39.09%	61.28%	0.4773
1-4-7	52.76%	61.11%	0.5663	77.94%	51.26%	0.6185	44.36%	63.14%	0.5211
1-5-6	38.61%	67.36%	0.4909	73.14%	54.76%	0.6263	31.89%	67.51%	0.4332
1-5-7	49.64%	67.21%	0.5710	81.53%	53.21%	0.6439	41.01%	67.86%	0.5112
2-4-6	47.24%	63.14%	0.5405	72.90%	51.61%	0.6044	38.13%	62.60%	0.4739
2-4-7	53.00%	61.39%	0.5689	77.46%	52.10%	0.6230	45.08%	63.51%	0.5273
2-5-6	37.89%	68.10%	0.4869	75.06%	57.01%	0.6480	33.09%	72.63%	0.4547
2-5-7	50.60%	69.41%	0.5853	82.73%	53.82%	0.6522	38.85%	69.83%	0.4992
3-4-6	44.12%	63.67%	0.5212	74.34%	52.90%	0.6181	38.37%	66.39%	0.4863
3-4-7	50.36%	61.58%	0.5541	76.50%	51.29%	0.6141	43.17%	64.98%	0.5187
3-5-6	35.01%	70.19%	0.4672	74.58%	56.65%	0.6439	29.98%	73.53%	0.4259
3-5-7	46.04%	68.82%	0.5517	83.21%	53.88%	0.6541	37.89%	70.22%	0.4922
Mean*	46.04%	65.25%	0.5364	76.66%	53.32%	0.6284	38.41%	66.96%	0.4851
Stand. dev.*	0.0599	0.0358	0.0377	0.0407	0.0200	0.0197	0.0478	0.0401	0.0336

*The mean and standard deviation values for the columns were incorporated on request of a reviewer

for other information retrieval tasks where the total number of relevant documents can be unknown. A typical user of NewsRec would read (and rate) a bunch of articles, train a classifier on the rated articles, obtain recommendations, and read (and rate) the next bunch of articles, train an updated classifier on all rated articles,.... If the true class labels of all articles are known, usage can be evaluated as follows: train a classifier on the first bunch of articles (training sets assigned by feedback option), compare predicted class labels with the true labels of the next bunch, train a new classifier on the first two bunchs of articles (training sets assigned by feedback option), evaluate it on the third bunch,.... Recall and precision for all evaluated bunchs are microaveraged; F1 is calculated on microaveraged recall and precision. On the examined website, about 50 new articles are presented per day. Thus, we set the size of a bunch to 50. Articles were sorted according to their order of appearance; the oldest articles were assigned to bunch 1.

6 Empirical Results

NewsRec had to be extented to allow for the comparison of explicit, implicit, and hybrid feedback options. An IT professional had to read and explicitly rate 1,185 articles of the Heise newsticker (http://www.heise.de/ct). In addition to the rating, the user had to submit whether (s)he would have requested the article during regular usage. This approach allowed us to evaluate the various feedback options on the same dataset, thus leading to comparable results. For explicit feedback, the true rating was used to determine the set of training articles. For implicit feedback, articles that would have been requested under regular usage were assigned to the set of interesting training documents, the others to the set of uninteresting training documents. For hybrid feedback, articles that would have been requested under regular usage and rated interesting were assigned to the set of interesting training documents, the remaining articles were assigned to the set of uninteresting training documents (assuming that requested but uninteresting articles were rated uninteresting). In order to prevent self fulfilling prophecies, no recommendations were given during data collection.

In total, 449 articles were interesting for the user (37%). Four hundred and twenty-five out of the 449 interesting articles would have been requested during regular usage (recall: 94%). Two hundred and twenty-five articles that were requested turned out to be uninteresting (precision: 65%). These results confirm our assumption that in news recommendation situations it should be more common that a user looks at an uninteresting article than that (s)he misses an interesting one.

Results for the examined feedback options are printed in Table 2. As we know, implicit feedback considers more articles interesting than explicit feedback. This leads to an increased recall at the expense of a lower precision. Hybrid feedback still considers not rated but interesting articles as uninteresting. The expectation to obtain higher precision but lower recall for hybrid feedback was confirmed by our empirical results.

In terms of F1, implicit feedback won the competition on this dataset due to the high recall values obtained.

For our application, however, precision is the more interesting measure. Here, hybrid feedback is best while the implicit counterpart only gets the last position.

Another possible cause for the results could be that the sets of training documents obtained through explicit feedback may lead to overfitted classifiers whereas hybrid feedback reminds the user that (s)he should rate a requested but uninteresting article as uninteresting.

7 Conclusions

Explicit feedback is considered as most reliable feedback option, but it is unpopular due to the required additional effort to rate all requested documents. Implicit feedback requires no additional effort at all, but it is considered as unreliable because it mislabels requested articles that are uninteresting and unrequested ones that are interesting. We combined both approaches within the so-called hybrid feedback option in order to reduce the burden of explicit feedback and lower the number of mislabeled training documents obtained by implicit feedback.

Obviously, implicit feedback leads to the largest absolute and relative numbers of positive training documents. The classifiers react accordingly. Implicit feedback leads to the highest recall, followed by explicit and hybrid feedback for our data. However, in terms of precision, the order is reversed for the same reasons.

The increase of recall for implicit feedback can outweigh the decrease in precision (as in the underlying example) as far as highest F1 values are concerned.

With respect to our application, hybrid feedback turns out to be the best choice: users, if in doubt, tend to look at articles. Thus, precision is the most important measure and hybrid feedback, which optimizes this measure, requires less effort than explicit feedback.

References

BOMHARDT, C. (2004): Newsrec, a SVM-driven personal recommendation system for news websites. In *WI 04: Proceedings of the IEEE/WIC/ACM International Conference on Web Intelligence*, 545–548.

CHEN, L. and SYCARA, K. (1998): Webmate: A personal agent for browsing and searching. In *Proceedings of the 2nd International Conference on Autonomous Agents and Multi Agent Systems, AGENTS 98*, 132–139.

CLAYPOOL, M., LE, P., WASED, M., and BROWN, D. (2001): Implicit interest indicators. In *IUI 01: Proceedings of the 6th international conference on Intelligent user interfaces*, 33–40.

JOACHIMS, T. (2002): Learning to classify text using support vector machines. Kluwer, Dordrecht.

JOACHIMS, T., FREITAG, D., and MITCHELL, T. (1997): WebWatcher: A tour guide for the world wide web. In *Proceedings of IJCAI97*, 770–775.

KELLY, D. and BELKIN, N. (2004): Display time as implicit feedback: understanding task effects. In *SIGIR 04: Proceedings of the 27th annual international ACM SIGIR conference on Research and development in information retrieval*, 377–384.

KELLY, D. and TEEVAN, J. (2003): Implicit feedback for inferring user preference: a bibliography. *SIGIR Forum, 37(2):*18–28.

KUSHMERICK, N., WELD, D., and DOORENBOS, R. (1997): Wrapper induction for information extraction. In *Intl. Joint Conference on Artificial Intelligence (IJCAI)*, 729–737.

LANG, K. (1995): NewsWeeder: Learning to filter netnews. *ICML*, 331–339.

MIDDLETON, S. (2001): Interface agents: A review of the field. *Technical Report ECSTR-IAM01-001, University of Southampton.*

MLADENIĆ, D. (2001): Using text learning to help web browsing. In *Proceedings of the 9th International Conference on Human–Computer Interaction.*

MOBASHER, B., COOLEY, R., and SRIVASTAVA, J. (2000): Automatic personalization based on web usage mining. *Communications of the ACM, 43(8)*, 142–151.

Classification in Marketing Science

S. Scholz and R. Wagner

Abstract This paper explores the role of clustering and classification methods in marketing beyond the prominent application of market segmentation (cf. Data analysis and decision support, Springer, Berlin, 157–176) Our study is based on a sample of more than 1,900 articles chosen from international peer-reviewed marketing journals. We apply a growing bisecting k-means for text classification to investigate the relation between the use of classification methods and the implications for scholars and practitioners revealed in the articles under consideration. Moreover, we evaluate the dependency between the use of classification techniques and the quality ratings of the papers.

This study highlights the application gaps of classification and clustering with respect to different areas of marketing science and, therefore, pinpoints domains of marketing science lacking a strong empirical foundation. Additionally, we draw implications for academic curricula.

1 Introduction

Pawlak (1982) claims that knowledge by itself is deep-seated in the classificatory abilities of human beings, in a basic but fundamental manner. Referring to the GfKl homepage, we define classification research as a set of embracing methods for organizing, structuring, and analyzing data. Not only classification, but various disciplines have contributed methodologies or substantial theories to marketing science. Mostly, these contributions evolve to conference series and serve as a nucleus for academic journals (e.g., Informs' "Marketing Science", reflecting the impact of operations research, or the "Journal of Consumer Psychology"). Although the importance of both basic and sophisticated classification is emphasized by

R. Wagner(✉)
DMCC Dialog Marketing Competence Center, University of Kassel, 34125 Kassel, Germany,
E-mail: rwagner@wirtschaft.uni-kassel.de

A. Okada et al. (eds.), *Cooperation in Classification and Data Analysis*,
Studies in Classification, Data Analysis, and Knowledge Organizations.
DOI: 10.1007/978-3-642-00668-5_10,

marketing scholars (e.g., Wedel and Kamakura 2003; Dolnicar et al. 2005), there is no particular journal focussing on classification challenges within the marketing discipline. The role of classification might be more general or subtle, or even nebulous. Thus, the aim of this paper is twofold:

1. To identify areas of marketing research that apply classification research
2. To assess the impact of published papers, which refer to sophisticated classification methods

The remainder of this paper is structured as follows: Next, we describe our data example by means of a text corpus and data preprocessing. In the third section, we introduce the text mining methodology, particularly the growing bisecting k-means used to identify distinctive topics discussed in the articles. The application is illustrated and selected results are presented in the fourth section. The paper concludes with a summary of the most important results and an outline of implications for academic education and research.

2 Data Description and Preprocessing

For our analysis we made up a text corpus of 1,965 articles taken from 13 marketing journals outlined in Table 1, for the time period 1998–2004. To evaluate the quality of the articles, we annotated the journal quality ratings provided by Harzing (2004). These ratings range from "highest quality" (equal to 1) to "lowest quality" (equal to 5).

Table 1 Structure of the available document collection

Title	Number of articles	Volume	Rating
Academy of Marketing Science Review (1)	41	1998–2004	3
Journal of Business Research (2)	517	1998–2004	2
Journal of Empirical Generalisations in Marketing Science (3)	7	1998–2001	–
International Journal of Research in Marketing (4)	129	1998–2004	2
Industrial Marketing Management (5)	297	1998–2004	3
Journal of the Academy of Marketing Science (6)	212	1998–2004	2
Journal of Advertising Research (7)	58	2000, 2003–2004	2
Journal of Market-Focused Management (8)	61	1998–1999, 2002	4
Journal of Retailing (9)	104	1999–2004	2
Marketing Letters (10)	155	1998–2004	2
Marketing Science (11)	97	2001–2004	1
Journal of Product Innovation Management (12)	63	1998–2000	2
Journal of Marketing Research (13)	224	1998–2004	1

All the articles of our marketing text corpus were downloaded in digital formats, typically as pdf-files. Each document was then converted into a standard ASCII text file. We conducted several steps of preprocessing the textual data: First, we removed all stop words from the text. The remaining words were reduced to the roots using Porter's (1980) stemming algorithm. We rely on a vector-space representation, in which each document $i \in D$ is weighted by means of the normalized TF-IDF scheme (Salton and Buckley 1988). In line with Dhillon and Modha (2001), we reduced the vocabulary size by cutting out terms that occurred in no more than 0.2% of the documents or appeared in over 15% of all documents in the text collection. Finally, we applied random mapping to reduce dimensionality of the data to 1,000. Random mapping requires less computational efforts but provides us with precision similar to latent semantic indexing (Bingham and Mannila 2001).

3 Growing Bisecting k-Means: A Methodological Outline

In order to find a meaningful partition of the text corpus, we use a variant of k-means clustering. The growing bisecting k-means approach bears a strong resemblance to the G-means clustering algorithm proposed by Hamerly and Elkan (2003) and the growing k-means algorithm outlined by Wagner et al. (2005) with respect to the challenge of market segmentation. The algorithm used herein is based on the classical bisecting k-means (Steinbach et al. 2000). It has two important advantages in comparison with standard k-means: First, the computational complexity is linear in the number of documents, since not each data point has to be compared with every cluster centroid. Second, the bisecting k-means does not necessarily tend to partition the data set in equally-sized clusters.

The bisecting-k-means algorithm proceeds in the following 5 steps: First, all document vectors d_i are assigned to one initial topic cluster (step 1). A cluster to split is chosen (step 2). We applied the simple rule to select the largest cluster for splitting, which leads to reasonable results in text mining applications (Steinbach et al. 2000). Splitting a cluster consists of applying the basic k-means algorithm several times with $k = 2$ (step 3) and keeping the best partition by means of an adequate criterion function, such as the average element-cluster similarity (step 4). The steps 2–4 have to be repeated until the desired number of clusters is reached (step 5).

The growing bisecting k-means algorithm extends the above classical bisecting k-means by autonomously determining the adequate number of clusters in a data set, which is recognized as one of the most difficult and largely unsolved problems of applying cluster analysis in marketing research (Dolnicar et al. 2005; Wagner et al. 2005). The procedure of the growing bisecting k-means is described in the pseudo-code outlined in Table 2.

The rationale of our approach is given by the assumption that document finger prints (made up by random mapping) of all documents assigned to a particular topic follow a multivariate normal distribution around the centroid that summarizes the

Table 2 Growing bisecting k-means algorithm

Step 1:	Assign all document vectors to one initial topic cluster
Step 2:	Apply bisecting k-means to the document collection to generate a k cluster solution
Step 3:	Test whether the documents assigned to a cluster follow a (multivariate) normal distribution
Step 4:	If the data meet the predetermined criteria of multivariate normal distribution, stop with a k^* cluster solution, otherwise increase k by one and return to step 1.

core content of this topic. Consequently, the major part of the assigned documents should also comprise this information. Those documents that also integrate additional (rather off-topic) issues are likely to mix in various different aspects that are only loosely connected to the core content represented by the centroid. Thus, we conclude that the density allocation of assigned documents decreases when moving away from the centroid in the vector-space representation. Another important factor is that the clusters received by the k-means algorithm follow a unimodal distribution in each dimension (Bock 1996; Ding et al. 2002). Accordingly, if the clusters are at least approximately multivariate normally distributed, the corresponding partition and its centroids are a good description of the underlying data structure.

Testing for multivariate normality to control the growing process could be done using Mardia's Test on Multinormality (e.g., Mecklin and Mundfrom 2004). However, this test requires great computational efforts in high dimensional text data because all dimensions of the input data need to be taken into account. Thus, we use the projection method described subsequently, which bears a resemblance to the approach outlined by Ding et al. (2002) and enables to project multidimensional data sets on one dimension. This facilitates the application of one-dimensional normality tests such as the powerful Anderson–Darling test (Stephens 1974). To check the most important directions of projection in the document space, we successively project all n_l document vectors assigned to each cluster onto those lines which are defined by the vectors $v_{ll'}$ that connect the corresponding cluster centroid C_l with the centroid of another cluster $C_{l'}$ simultaneously (see Fig. 1 for graphical illustration in a two-dimensional data space):

$$v_{ll'} = C_l + \beta(C_l - C_{l'}) \qquad \forall l,\ l' = 1, \ldots, k \text{ and } l \neq l'. \tag{1}$$

This projection leads to a serious reduction of computational efforts when testing for normality. All directions of the document space that are most important for clustering are considered (Ding et al. 2002). Using the Anderson–Darling test, we can check whether the documents assigned to a cluster are normally distributed with respect to each projection $v_{ll'}$ or not. We assume that clusters are approximately multivariate normally distributed when a cluster follows a normal distribution for all projections with $l' = \{1, \ldots, k \backslash l\}$. As various clusters and various directions for each cluster have to be checked, we use Bonferroni adjustment when selecting the significance level α. The algorithm stops at the smallest partition of the data set,

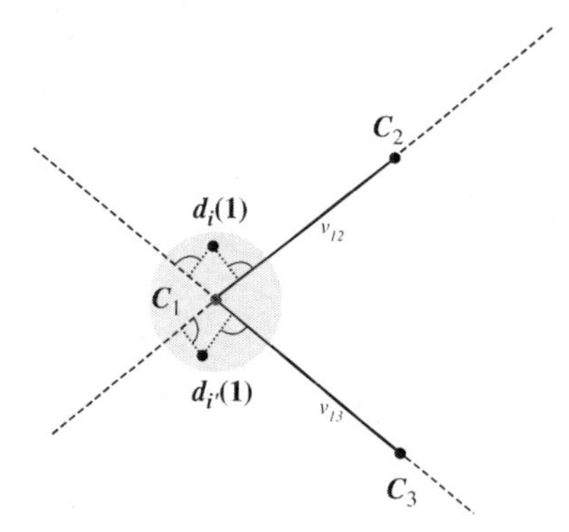

Fig. 1 Directions of projection of cluster C_1 in a two-dimensional document space with three clusters

in which all clusters are normally distributed for all possible projections: This partition is regarded as the most appropriate segmentation of the document collection and determines the number of clusters given in the document space.

4 Application to Marketing Text Corpus

To determine an adequate critical value for the above Anderson–Darling test, we apply a Bonferroni correction factor of 100 in order to reduce the probability of Type 1 errors. The corresponding critical value equals 1.692. When applying the growing bisecting k-means algorithm to our marketing text corpus, the algorithm stops with $k^* = 35$ resulting in a maximum test value of 1.20 for all projections in each cluster. Thus, we assume that this partition is an adequate representation of the structure of the data set. Thereby, cluster size ranges from 2 to 478, confirming that the growing bisecting k-means does not have the tendency to produce equally-sized clusters.

We annotated the medoid document to each cluster as a proxy for the topics discussed in the respective clusters. (Detailed cluster explanations are skipped because of space restrictions.)

Table 3 provides an overview of the 35-cluster solution and a short description of the main topic of each cluster. For this study, the impact of classification methods on research quality appears to be most interesting.

In order to analyze the impact of classification research in the marketing literature, we implemented two rough measures, by which we are able to appraise (1) the impact, or more precisely, the weighted arithmetic mean of the journal quality rating, and (2) the degree of application of classification methods in the marketing text corpus.

For the appraisal of the quality/impact of the topic clusters (1), we annotated the average journal quality rating to each cluster of the partition provided by the growing bisecting k-means (see column 3 in Table 3). In doing so, we have been able to establish a proxy measure for the impact of the topics primarily discussed in the 35 clusters on the marketing research community. To analyze the degree of discussion

Table 3 Overview of topics discussed in 35-cluster partition of the text collection

No.	Title	Size	Rating	$\overline{Freq}$	Journals
1	Corporate social responsibility	2	2.00	1.5	4, 13
2	Franchising	4	2.00	5.5	2,9
3	E-commerce	6	2.50	4.33	4, 5, 10
4	Ethic in marketing	14	2.14	6.22	1, 2, 5, 6, 9,
5	New product development	23	2.26	3.5	2, 5, 6, 12
6	Miscellaneous	478	2.34	4.5	1–13
7	Difference between men and women	17	2.47	2.88	1, 2, 4, 5, 7, 9, 10
8	Export	17	2.35	4.94	2, 4, 5, 7, 13
9	Health care	10	2.00	2.7	2
10	Price matching	5	2.00	6.8	1, 2, 9, 11, 13
11	Market entry	9	2.00	3	2, 4, 12, 13
12	Selling, salespersons, sales forces	61	2.44	2.23	1, 2, 4, 5, 6, 8–13
13	Consumer and coupons	9	2.00	4.67	2, 9, 10, 13
14	Asia, China, cross-cultural	34	2.26	1.62	2, 4–9, 12, 13
15	SERVQUAL	90	2.13	3.81	1, 2, 4–10, 12, 13
16	Conjoint, consumer behavior	128	2.00	13.3	2, 4–13
17	Differences towards culture and gender	130	2.10	13.62	1, 2–7, 9–13
18	Mathematic models in marketing	257	1.91	17.07	1–11. 13
19	Market orientation	43	2.84	2.07	1, 2, 4–6, 8, 13
20	International business, expansion	79	2.22	5.33	1, 2, 4–13
21	Service and consumer satisfaction	18	2.17	2.22	1, 2, 6, 8–10, 13
22	Consumer orientation	87	2.20	4.03	1–8, 10, 11, 13
23	Foreign direct investment	13	1.92	3.62	2, 10, 11, 13
24	Organizational theory and measurement	37	2.38	3.81	1, 2, 5–8, 10, 12, 13
25	Family, influence of parents	20	2.10	2.35	2–4, 6, 9, 10, 13
26	Supply chain management	103	2.22	4.2	1, 2, 4–13
27	Business networks	62	2.31	6.53	2, 4–13
28	Shopping environment	85	2.00	5.14	1, 2, 4–7, 9–11, 13
29	Tourism	14	2.00	1.29	2, 4, 9
30	Seller and reseller	9	2.67	2	2, 5, 6
31	Counter-trade	3	3.33	3	5, 8
32	Advertising agencies	23	2.09	5.87	2, 4, 5, 7, 12, 13
33	New product introduction	24	1.96	4.75	1, 2, 4–7, 9, 11–13
34	Hedonic consumption	40	2.18	1.53	1, 2, 4–13
35	External reference prices	11	2.27	2.73	2, 5, 6, 9, 10, 13

or application of clustering and classification methods in the 35 clusters (2), we extracted a subset of the bag-of-words model that comprises 30 terms, which are highly related to methods in classification research. These terms include, amongst others, names of methods discussed in classification research, such as finite mixture models, fuzzy sets, and principal component analysis. Moreover, the surnames of prominent researchers in the scientific classification community, such as Bock, Gordon, and Gower were incorporated in the extracted terms. We calculated the average frequency of usage of these 30 terms $\overline{freq}$ in each cluster. By means of these rough measures, we checked the relationship between the average quality rating and the use of terms associated with classification research. As indicated in Table 3, cluster 1, which deals explicitly with quantitative models in marketing and, therefore, by definition include sophisticated methods of classification, yields both the best quality rating and the highest use of words associated with classification research. Moreover, clusters 2 and 3, which also yield high scores for the variable $\overline{freq}$, refer directly to the frequent use of classification techniques. However, the lower score of cluster 3 might be a result of combining sophisticated classification with specialized techniques of preference measurement. Those clusters that do not frequently use terms associated with classification methods are listed near the bottom of Table 1. An example is cluster 30, which deals with the measurement of market orientation. Although these articles deal with empirical measurement, the authors do not employ sophisticated classification techniques, i.e., taking heterogeneity into account. This deficit explains the low quality rating of this cluster.

Across clusters, the Spearman rank correlation ($r = 0.35$) reveals a highly significant ($p < 0.01$) and positive relation between the usage of classification methods and the quality ranking of the papers. Given the wide range of factors influencing the quality of a paper, we assert that the impact of classification research is substantial for the quality of academic research in the domain of marketing. Taking a more detailed look, the following pattern emerges: the more specialized the methods, the higher the impact of the articles comprised in the respective clusters. In line with this rule-of-thumb, referring to Professor Bock has a significant impact on the weighted arithmetic mean of the quality rating of the clusters ($r = 0.28$ with $p < 0.05$) while referencing to Professor Gower (his famous Gower coefficient is frequently used in marketing research) has no significant effect.

5 Conclusions

This paper seeks to identify areas in marketing research that apply classification methods. By applying a growing bisecting k-means algorithm to a text corpus of 1,965 academic articles resulted in 35 clusters reflecting particular topics within the domain of marketing science. This study reveals a significant and positive relation between the use of sophisticated classification methods and the quality assessment of the research output. Additionally our results explain the lower quality assessments of particular topics within marketing research. Moreover, the application of highly specialized methods is a characteristic feature of high quality papers.

References

BINGHAM, E. and MANNILA, H. (2001): Random Projection in Dimensionality Reduction: Applications to Image and Text Data. In: F. Provost and R. Srikant (Eds.) *Proceedings of the 7th ACM SIGKDD International Conference on Knowledge Discovery and Data Mining*, 245–250.

BOCK, H.H. (1996): Probability Models in Partitional Cluster Analysis. *Computational Statistics and Data Analysis, 23(5)*, 5–28.

DHILLON, I. and MODHA, D. (2001): Concept Decompositions for Large Sparse Text Data Using Clustering. *Machine Learning, 42(1)*, 143–175.

DING, C., HE, X., ZHA, H., and SIMON, H. (2002): Adaptive Dimension Reduction for Clustering High Dimensional Data. In: *Proceedings of the 2nd IEEE International Conference on Data Mining*.

DOLNICAR, S., FREITAG, R., and RANDLE, M. (2005): To Segment or Not to Segment? An Investigation of Segmentation Strategy Success Under Varying Market Conditions. *Australasian Marketing Journal, 13(1)*, 20–35.

HAMERLY, G. and ELKAN, C. (2003): Learning the k in k-Means. In: *Advances in Neural Information Processing Systems, Vol. 17*. MIT, Cambridge, MA.

HARZING, A.W. (2004): Journal Quality List. University of Melbourne. http:\\www.harzing.com, 2004.

MECKLIN, C.J. and MUNDFROM, D.J. (2004): An Appraisal and Bibliography of Tests for Multivariate Normality. *International Statistical Review, 72*, 123–138.

PAWLAK, Z. (1982): *Rough Sets: Theoretical Aspects of Reasoning about Data*. Kluwer, Dordrecht.

PORTER, M.F. (1980): An Algorithm for Suffix Stripping. *Program, 14(3)*, 130–137.

SALTON, G. and BUCKLEY, C. (1988): Term-Weighting Approaches in Automatic Text Retrieval. *Information Processing and Management, 24(5)*, 513–523.

STEINBACH, M., KARYPIS, G., and KUMAR, V. (2000): A Comparison of Document Clustering Techniques. In: *KDD Workshop on Text Mining*.

STEPHENS, M. (1974): EDF Statistics for Goodness of Fit and Some Comparisons. *Journal of the American Statistical Association, 69*, 730–737.

WAGNER, R., SCHOLZ, S.W., and DECKER, R. (2005): The Number of Clusters in Market Segmentation. In: D. Baier, R. Decker, and L. Schmidt-Thieme (Eds.): *Data Analysis and Decision Support*. Springer, Berlin, 157–176.

WEDEL, M. and KAMAKURA, W.A. (2003): *Market Segmentation: Conceptual and Methodological Foundations* (2nd edn.). Springer, New York.

Deriving a Statistical Model for the Prediction of Spiralling in BTA Deep-Hole-Drilling from a Physical Model

C. Weihs, N. Raabe, and O. Webber

Abstract One serious problem in deep-hole drilling is the occurrence of a dynamic disturbances called spiralling. A common explanation for the occurrence of spiralling is the coincidence of time varying bending eigenfrequencies of the tool with multiples of the spindle rotation frequency. We propose a statistical model for the estimation of the eigenfrequencies derived from a physical model. The major advantage of the statistical model is that it allows to estimate the parameters of the physical model directly from data measured during the process. This represents an efficient way of detecting situations in which spiralling is likely and of deriving countermeasures.

1 Introduction

Deep hole drilling methods are used for producing holes with a high length to diameter ratio, good surface finish and straightness. For drilling holes with a diameter of 20 mm and above, the BTA deep hole machining principle is usually employed (VDI 1974). The necessarily slender tools, consisting of a boring bar and head, have low dynamic stiffness properties. Therefore deep-hole-drilling processes are at a high risk of dynamic disturbances such as spiralling, which causes a multi-lobe-shaped deviation of the cross section of the hole from absolute roundness, see Fig. 1.

As the deep hole drilling process is often applied during the last production phases of expensive workpieces, process reliability is of prime importance. Prediction and prevention of spiralling are therefore highly desirable.

By using a finite elements model to determine drilling depth dependent bending eigenfrequencies of the tool, spiralling was shown to reproducibly occur when one of its slowly varying eigenfrequencies intersects with an uneven multiple of the tool rotational frequency (Gessesse et al. 1994). This suggests preventing spiralling by

C. Weihs(✉)
Chair of Computational Statistics, University of Dortmund, Dortmund, Germany,
E-mail: claus.weihs@T-Online.de

A. Okada et al. (eds.), *Cooperation in Classification and Data Analysis*,
Studies in Classification, Data Analysis, and Knowledge Organizations.
DOI: 10.1007/978-3-642-00668-5_11,

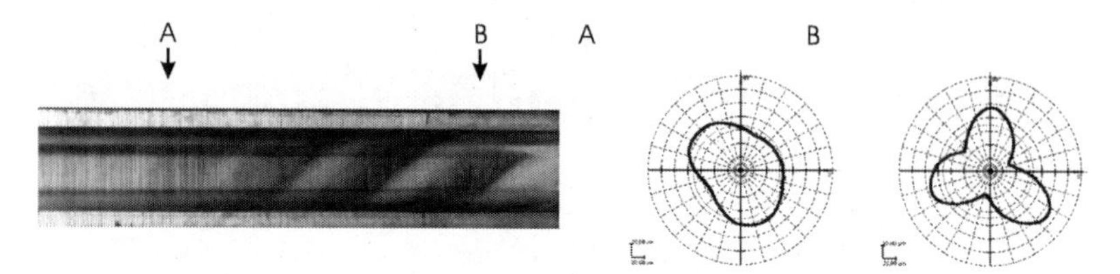

Fig. 1 *Left:* Longitudinal section of a bore hole showing marks resulting from spiralling. *Right:* Associated roundness charts

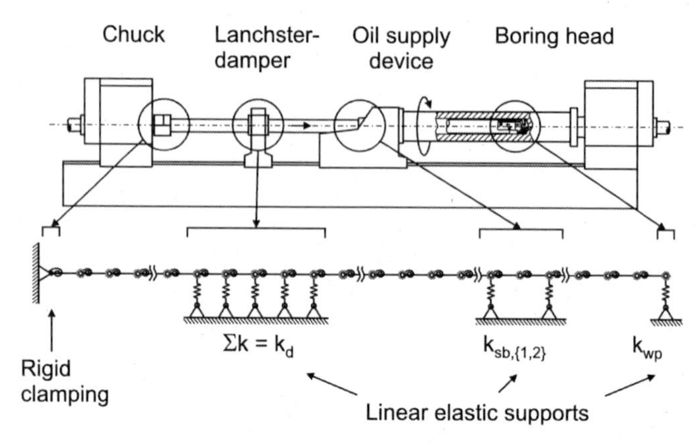

Fig. 2 Experimental setup (*top*) and proposed modelling approach (*bottom*)

avoiding these critical situations. Unfortunately the practical application of the finite elements model is limited as it has to be calibrated using experimentally determined eigenfrequencies.

Earlier investigations demonstrated that the courses of the bending eigenfrequencies clearly show in spectrograms of the structure borne sound of the boring bar, which can be recorded during the process (Raabe et al. 2004). In this paper this signal is used to statistically estimate the parameters of a physical model of the bending eigenfrequencies. A lumped mass model is used to calculate the tools bending eigenfrequencies. It includes the physical parameters of the process allowing to directly calculate the influences of their variations on the eigenfrequency courses. However, this model contains some unknown parameters and naturally the measurement is subject to random error. It is therefore combined with a statistical model allowing the estimation of the unknown parameters by the Maximum Likelihood method.

The work presented in this paper is based on experiments carried out on a CNC deep hole drilling machine type Giana GGB 560 (see Szepannek et al. 2006 for technical details). Self excited torsional vibrations were prevented through the application of a Lanchester-damper. The damper was moved at feed speed together with the boring bar, implying a constant axial position of the damper relative to the tool. In order to detect bending vibrations occurring during the process, time series of the lateral acceleration of the boring bar were recorded. The experiments were carried out with stationary tool and rotating workpiece. The experimental setup is illustrated in Fig. 2.

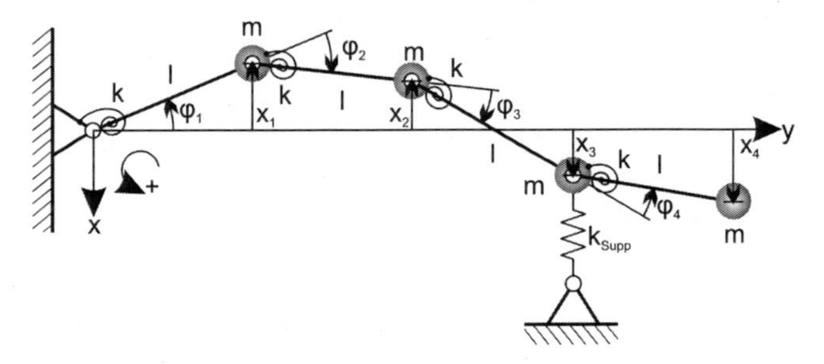

Fig. 3 Detailed modeling principle: Regular linear elastic chain with additional linear elastic support

2 Physical Model

For formulating the model the BTA system was reduced to its most important components. These are the tool, the Lanchester damper, two oil seal rings within the oil supply device and the workpiece, again see Fig. 2, top. Under operating conditions, the latter components act as lateral elastic constraints of the boring bar. While the damper stays in the same location relative to the boring bar, the oil supply device is kept at constant distance relative to the workpiece and therefore moves at feed speed relative to the boring bar during the process. The workpiece permanently acts on the tip of the tool.

As illustrated in Fig. 3 by an exemplary system with four degrees of freedom, the bar is subdivided into N elements of identical length l for constructing the lumped mass model. These elements are linked to form a regular linear elastic chain comprising N identically spaced and elastically linked masses. Additional linear elastic supports represent the constraints resulting from the supporting elements. Adopting the x-coordinates as generalized coordinates and assuming only small deflections we can write the homogenous equations of motion of the system as

$$[M]\{\ddot{x}\} + [K(l_B)]\{x\} = \{0\} \quad \text{with} \quad [K(l_B)] = [K_{\text{Tool}}] + [K_{\text{Supp}}(l_B)], \tag{1}$$

where $[M]_{N\times N}$ and $[K(l_B)]_{N\times N}$ are the mass and stiffness matrices of the system and l_B represents the actual drilling depth. The stiffness-matrix can be decomposed into the stiffness matrix $[K_{\text{Tool}}]$ of the boring bar and a matrix $[K_{\text{Supp}}(l_B)]$ containing the stiffness influences of the supporting elements. $[K_{\text{Tool}}]$ is time constant and can be computed from the physical and geometrical properties of the tool (see Szepannek et al. 2006), whereas $[K_{\text{Supp}}(l_B)]$ changes stepwise with increasing drilling depth due to the movement of the oil supply device relative to the boring bar. Furthermore, $[K_{\text{Supp}}(l_B)]$ generally is unknown. More precisely, all elements of $[K_{\text{Supp}}(l_B)]$ are zero except of these elements on the main diagonal that correspond to a supporting element in the setup. The values of these matrix entries are the unknown parameters of the model.

The stiffness influences of the workpiece and the two seals of the stuffing box within the oil supply device are each assumed to act pointwise and are therefore modelled by one single parameter each (k_{wp}, $k_{sb\{1,2\}}$). The Lanchester-damper contacts the boring bar within a region of nominal length l_d. It is assumed, that this region may be reduced, e. g. by wear. So two parameters $\delta l_{d,r}$ and $\delta l_{d,l}$ representing a right- and left-hand truncation of l_d are added. The stiffness influence of the damper (k_d) is equally distributed over the elements within the remaining region of length $l_d - \delta l_{d,r} - \delta l_{d,l}$.

The stiffness constants k_{wp}, $k_{sb\{1,2\}}$, k_d together with $\delta l_{d,r}$, $\delta l_{d,l}$, which define the matrix $[K_{\text{Supp}}(l_B)]$, are a priori unknown and cannot be measured directly. These parameters therefore have to be estimated. For calculating the eigenfrequencies from the model the homogeneous equations of motion of the system (see above) have to be solved for each regarded value of the drilling depth l_B. This leads to the following eigenvalue-problem

$$\left([K(l_B)] - \omega^2[M]\right)\{x\}e^{I\omega t}. \tag{2}$$

The solution of this problem consists of the eigenvalues ω^2_{r,l_B}, the N squared eigenfrequencies of the model, and the eigenvectors $\{\Psi\}_{r,l_B}$, the corresponding N mode shape vectors.

3 Statistical Model

For the estimation of the unknown parameters a statistical approach using the already introduced structure borne sound is proposed. In the following the data measured in a location between damper and oil supply device is exemplarily used.

To provide a basis for statistical estimation of the unknown parameters p, the following statistical model is proposed

$$S_k(\omega, l_B; p) = \left|\alpha_{jk}(\omega, l_B; p)\right|^2 \cdot \left|\alpha_j^*(\omega)\right|^2 \cdot S_\epsilon(l_B). \tag{3}$$

For each value of the hole depth l_B the term $S_k(\omega, l_B; p)$ presents the periodogram of the structure borne sound measured at a location corresponding to element k. Due to the discreteness of the physical model l_B changes stepwise and so the periodograms are computed based on non-overlapping time-windows. The model writes these periodograms $S_k(\omega, l_B; p)$ as the product of a systematic component $\left|\alpha_{jk}(!, l_B; p)\right|^2 \cdot \left|\alpha_j^*(\omega)\right|^2$ (the spectral density of the process) and a stochastic exciting component $S_\epsilon(l_B)$, the periodogram of a white noise process. The systematic component consists of the frequency response function (FRF) series $\alpha_{jk}(\omega, l_B; p)$ and the time constant $\alpha_j^*(\omega)$, which transforms the white noise process into the excitation in element j. In a first attempt $\alpha_j^*(\omega)$ for each frequency ω is set to its mean observed amplitude value. Refinements like fitting $\alpha_j^*(\omega)$ and p alternately are imaginable in later investigations. For a better impression Fig. 4 gives a graphical representation of the proposed statistical model.

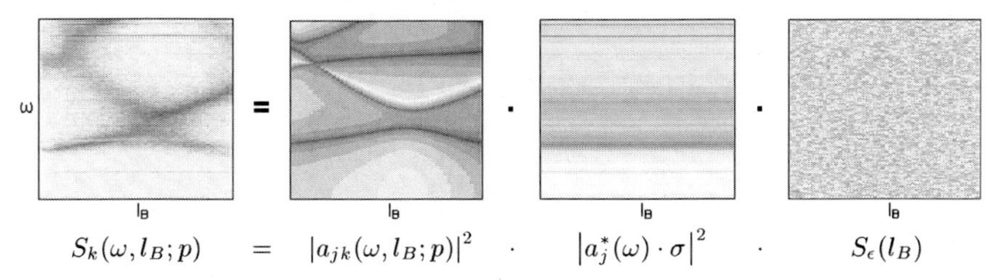

Fig. 4 Visualization of the statistical model

3.1 FRF Computation

For the computation of a FRF damping has to be included. The most straightforward way of doing this is assuming proportional damping, implying the damping matrix $[C(l_B)] = \beta[K(l_B)] + \gamma[M]$. This leads to the two further model parameters β and γ. Therefore the list of model parameters reads

$$p = (k_{wp}, k_{sb\{1,2\}}, k_d, \delta l_{d,r}, \delta l_{d,l}, \beta, \gamma). \tag{4}$$

Computation of the FRF necessitates the definition of the points of excitation j and response k. The excitation point j was chosen to be the last element N, because at this position the cutting process takes place. Element k naturally corresponds to the point at which the considered signal is recorded. The FRF can then be computed by

$$\alpha_{jk}(\omega, l_B; p) = \omega^2 \sum_{r=1}^{N} \frac{\Psi_{jr,l_B}\Psi_{kr,l_B}}{k_{rr,l_B} - \omega^2 m_{rr} + i\omega c_{rr,l_B}}, \tag{5}$$

where Ψ_{jr,l_B} denotes the j-th element of the r-th mode shape vector $\{\Psi\}_{r,l_B}$, and k_{rr}, m_{rr} and c_{rr} are the r-th diagonal elements of the modal stiffness-, mass- and damping matrices, respectively. These matrices can directly be derived from quadratic forms of the mode shape vectors and the stiffness- and mass matrices $[M]$ and $[K(l_B)]$ (Ewins 2000). Finally, the eigenfrequencies ω'_{r,l_B} of the proportionally damped system are given by

$$\omega'_{r,l_B} = \omega_{r,l_B}\sqrt{1 - \left(\beta\omega_{r,l_B}/2 + \gamma/\left[2\omega_{r,l_B}\right]\right)^2}. \tag{6}$$

As for this description once a specific p is chosen, the corresponding eigenfrequencies can be determined.

3.2 Maximum Likelihood Estimation

The parameters of the systematic model part can be estimated using the Maximum Likelihood method. The Likelihood-function can be derived by connecting the following well known results:

1. The periodogram $I_x(\lambda)$ of each stationary process X_t with a moving-average representation

$$X_t = \sum_{u=-\infty}^{\infty} \beta_u \epsilon_{t-u} \quad \text{with} \sum_{u=-\infty}^{\infty} (1 + |u|)|\beta_u| < \infty \tag{7}$$

implying the spectral density $f_x(\lambda) = \left|\sum_u \beta_u e^{i2\pi\lambda u}\right|^2$ has an exponential distribution at each Fourier frequency λ with parameter $1/f_x(\lambda)$. Then periodogram ordinates at different Fourier frequencies are asymptotically independent (Schlittgen and Streitberg 1999, p. 364).
2. Each stationary process X_t with continuous and for all λ non-negative spectral density $f_x(\lambda)$ has an infinite moving-average representation (Schlittgen and Streitberg 1999, p. 184).

Assumption 2 can be seen as fulfilled, as all inspected spectrograms clearly show values different from zero for all frequencies and time points. Assumption 2 substantially implies assumption 1, so for each Fourier frequency ω and hole depth l_B the distribution function of $S(\omega, l_B; p)$ is approximatively given by

$$d.f.(s) = f(\omega, l_B; p)^{-1} e^{f(\omega, l_B; p)^{-1} s}, \tag{8}$$

where $f(\omega, l_B; p) = \left|\alpha_{jk}(\omega, l_B; p)\right|^2 \cdot \left|\alpha_j^*(\omega)\right|^2$.

Using the asymptotical independence of periodogram ordinates of different frequencies and assuming independence for different hole depths, the Log-Likelihood-function is given by

$$LL(p) = \sum_{l_B} \sum_{\omega} \left[\ln \frac{1}{f(\omega, l_B; p)} - \frac{S(\omega, l_B; p)}{f(\omega, l_B; p)} \right]. \tag{9}$$

The ML-estimators are the set p_{ML} of parameters maximizing this function. With these parameters the estimated eigenfrequencies can be derived as illustrated in the last two sections.

The introduced model has been successfully fitted to different experiments by using the search-based method by Nelder and Mead (1965) for the maximization of the Log-Likelihood-function. Figure 5 shows an exemplary comparison between an acceleration spectrogram and the bending eigenfrequencies computed from the fitted model for a process without spiralling.

Even though the second and third eigenfrequency seem to over-estimate the area of elevated amplitudes the pattern in the spectrogram is clearly represented by the fit. So apart from possible model refinements, these results, which are similar for all other experiments investigated up to now, support the connection of the physical model with the statistical model as a basis for estimating its parameters from spectrogram data.

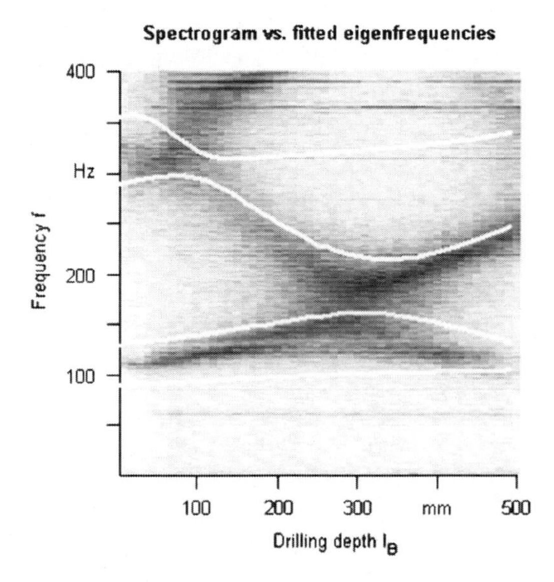

Fig. 5 Comparison between acceleration spectrogram and fitted eigenfrequencies

4 Summary and Outlook

The presented paper shows that a connection of a physical and a statistical model helps to estimate the bending eigenfrequencies of a deep-hole-drilling tool from data available during the process. Bending eigenfrequencies are known to cause spiralling when crossing multiples of the spindle rotational frequency. By supervising the estimated eigenfrequencies, shifts in the process dynamics can be detected and crucial situations can be predicted.

The supervision may be possible within a batch production, where after each completely drilled workpiece the eigenfrequencies are checked and the necessity of parameter changes is decided. In the actual form fitting the model is too time extensive to allow online intervention. But if the model can be simplified implying faster fitting procedures, strategies such as control charts for the eigenfrequencies could be feasible as well.

Simplifications of the physical model are possible by concentrating on the relevant regions of the spectra or modifications of the discretization. The statistical model may be simplified by estimating the spectra for frequency bands instead of Fourier frequencies using consistent estimates as introduced in Schlittgen and Streitberg (1999). Furthermore for a more efficient way of estimating the eigenfrequencies historical data may be used in connection with the physical model.

In future experiments roundness errors of the drilled workpieces will be measured at different equally spaced hole depth points. These measurements represent a quantization of the effect of spiralling over time and so help to investigate the development of spiralling more closely in different situations. Here main features of interest are whether spiralling starts rapidly or develops slowly and if the magnitude of spiralling depends on how quickly the frequency crossing takes place. As the measurement of roundness errors is not possible in production the potentials of estimating these from the spectrogram data will be checked as well.

Acknowledgements This work has been supported by the Collaborative Research Center "Reduction of Complexity in Multivariate Data Structures" (SFB 475) of the German Research Foundation (DFG).

References

EWINS, D. (2000): *Modal Testing: Theory, Practice and Application,* 2nd edn. Research studies press, Badlock.

GESSESSE, Y.B., LATINOVIC, V.N., and OSMAN, M.O.M. (1994): On the Problem of Spiralling in BTA Deep-Hole Machining. *Transaction of the ASME, Journal of Engineering for Industry* 116:161–165.

NELDER, J.A. and MEAD, R. (1965): A Simplex Method for Functional Minimization. *Computer Journal* 7:308–313.

RAABE, N., THEIS, W., and WEBBER, O. (2004): Spiralling in BTA Deep-Hole Drilling – How to Model Varying Frequencies. In *Conference CD of the Fourth Annual Meeting of ENBIS 2004,* Copenhagen.

SCHLITTGEN, R. and STREITBERG, B.H. (1999): *Zeitreihenanalyse*. Oldenbourg, München.

SZEPANNEK, G., RAABE, N., WEBBER, O., and WEIHS, C. (2006): Prediction of Spiralling in BTA Deep-Hole Drilling – Estimating the System'S Eigenfrequencies. *Technical Report,* SFB 475, Dortmund.

VDI (1974): VDI-Richtlinie 3210: Tiefbohrverfahren. *VDI Düsseldorf.*

Analyzing Protein–Protein Interaction with Variant Analysis

G. Ritter and M. Gallegos

Abstract Biochemistry teaches that the functioning of a biological organism is to a large extent determined by the interaction of biomolecules. One of these interactions is that of two proteins. Besides experimental methods, sequence information in combination with computational methods can be used to shed light in their mode of operation. For this purpose we present a method based on variant analysis, a general approach for treating ambiguities in so-called *ambiguous data sets*. After sketching an outline of variant analysis, we apply it together with a coin-tossing model to the problem of multiple local string alignment in protein sequences. The resulting new algorithm is shown to detect the target proteins, the recognition motif, and the association sites, also in a contaminated environment.

1 Introduction

Modern biology and biochemistry is, among other things, concerned with analyzing protein function and their interaction with polynucleotides and with other proteins. Besides the biological experiment such as phage display, SPOT analysis, NMR spectroscopy, fluorescence titration, statistical and computational methods based on the huge amount of sequence information available have been shown in the last 15 years to be effective tools for this purpose. To this end, advanced statistical and computational methods are employed. We mention in this respect algorithms such as the EM algorithm and Gibbs sampling, and statistical models such as the HMM. They allow to determine transcription factor binding sites and their motifs, see Lawrence and Reilly (1990), Lawrence et al. (1993), Liu (1994), Liu et al. (1995). Gallegos and Ritter (2006) showed that variant analysis, a general statistical theory for treating ambiguities, could be applied for this purpose, too. As an example we used the

G. Ritter(✉)
Faculty of Computer Science and Mathematics, University of Passau, 94030 Passau, Germany,
E-mail: ritter@fim.uni-passau.de

A. Okada et al. (eds.), *Cooperation in Classification and Data Analysis*,
Studies in Classification, Data Analysis, and Knowledge Organizations.
DOI: 10.1007/978-3-642-00668-5_12,

binding sites for the cyclic AMP receptor protein (CRP) in the promoters of 18 genes of *Escherichia coli*, Stormo and Hartzell (1989), a protein–DNA interaction. We show in this note that the method is also effective in the analysis of protein–protein interactions. As an example, we study the recognition sites of two GYF domain-containing proteins of yeast, SMY2 and SYH1, see Kofler et al. (2005).

2 Parameter Estimation Under Ambiguity and Contamination

Contrary to statistics, pattern recognition deals with objects such as stars, plants, images, or signals rather than with data vectors as input to an analysis. In general, feature vectors are extracted from these objects and subsequently analyzed by statistical or other methods. It often happens that the extraction depends on an interpretation of the object which may not be unique in this stage of analysis. In this case we face ambiguity, a phenomenon that occurs in particular if the features are extracted by a machine. In the case of several reasonable, but not necessarily correct, interpretations, one may extract several feature sets of each object, one for each reasonable interpretation. This means that each object may be represented by several distinct rows in the data set, called *variants* of the object, Ritter (2000). The variant that corresponds to the correct interpretation is the *regular* variant, the others are *irregular*. We call such a data set *ambiguous*, see Table 1. In some cases, some or all objects may possess more than one regular variant.

All questions about classical data sets may also be asked about ambiguous data sets – parameter estimation, discriminant analysis, clustering, ... An obstacle in the analysis of an ambiguous data set is the fact that it should be based on the regular variant of each object so that while carrying out the primary task the regular variant has to be estimated at the same time. This program was carried out for discriminant analysis, Ritter and Gallegos (2000), pure selection of the (or a) regular variant, Ritter and Gallegos (2002), and parameter estimation, Gallegos and Ritter (2006). Applications to image processing appear in Ritter and Pesch (2001), Ritter and Schreib (2000, 2001), Ritter and Gallegos (2000), and an application to motif discovery in genetics in Gallegos and Ritter (2006).

Table 1 Three-dimensional data sets, (a) classical and (b) ambiguous. Whereas the six observations in data set (a) belong to six different objects, the six observations in data set (b) are extracted from three objects. Only one of the two variants of object 1 and one of the three variants of object 2 is a valid representative of its object. All variants of the same object are at first equally important

(a)				(b)			
Obser_1	5.37	1.62	2.45	Object_1	2.37	3.62	4.41
Obser_2	4.11	2.21	2.13	Object_1	1.14	1.21	3.12
Obser_3	3.34	4.54	5.46	Object_2	3.30	5.62	7.33
Obser_4	8.35	6.76	7.78	Object_2	8.11	6.29	7.13
Obser_5	1.36	2.48	1.41	Object_2	3.11	4.21	3.13
Obser_6	5.76	7.61	2.15	Object_3	6.54	5.22	8.46

2.1 A Statistical Model of Variants

Let $x_i = (x_{i,1}, \ldots, x_{i,h_i}, \ldots, x_{i,b_i})$, $x_{i,k}$ in some sample space E, represent the b_i variants of object $i \in 1..n$. For the purpose of this note, E is finite. Let us distinguish between two cases:

1. At least one of the b_i variants of object i is regular, say h_i. We then assume that x_{i,h_i} is distributed according to some distribution with density f_γ for γ in some parameter space and call i a *regular object*.
2. None of the b_i variants is regular. In this case, object i is an *outlier*.

We assume that there are $r \leq n$ regular objects. The choice of the "correct" number is made later by validating the results obtained for various values of r. Central to variant analysis is the notion of a (variant) *selection* $h = (h_1, \ldots, h_n)$, $h_i \in 0..b_i$. The relation $h_i = 0$ specifies object i as an outlier, whereas $h_i > 0$ means that h_i is the site of its regular variant. Thus, a selection implicitly contains the information about the regular objects in the data set. For example, if $r = 2$, ([object_1, 1], [object_2, 3], [object_3, 0]) is a selection in the data set (b) of Table 1. This selection considers objects 1 and 2 as regular and object 3 as an outlier. Our main objective is estimating the "true" selection and, thereby, the parameter γ.

Denote the cross section $(x_{i,h_i})_{h_i > 0}$ specified by a selection $\mathbf{h}$ by $x_{\mathbf{h}}$; it is a classical data set with one row per regular object i. The cross section in Table 1 of the selection above is

object_1	2.37	3.62	4.41
object_2	3.11	4.21	3.13

Given i.i.d. random variables X_i, $1 \leq i \leq r$, distributed according to some unknown "true" distribution μ, the arithmetic means

$$\frac{1}{r}\sum_{i=1}^{r} -\ln f_\mu(X_i) \quad \text{and} \quad \frac{1}{r}\sum_{i=1}^{r} \ln \frac{f_\mu}{f_\gamma}(X_i)$$

converge to the entropy $-\mathrm{e}\ln f_\mu(X_1)$ of μ and to the Kullback–Leibler divergence $\mathrm{e}\ln \frac{f_\mu}{f_\gamma}(X_1)$ of μ and γ, respectively, P-a.s. Hence, given a finite sequence $x_1, \ldots, x_r$ of observations, the means

$$\frac{1}{r}\sum_{i=1}^{r} -\ln f_\mu(x_i) \quad \text{and} \quad \frac{1}{r}\sum_{i=1}^{r} \ln \frac{f_\mu}{f_\gamma}(x_i)$$

are sample versions of these quantities. Neither of the two can be computed since f_μ is unknown, but their sum $\frac{1}{r}\sum_{i=1}^{r} -\ln f_\gamma(x_i)$ is an expression of γ alone. Two desirable aims are small entropy and small Kullback–Leibler divergence. These aims can be simultaneously achieved by minimizing this sum over γ. In the context of irregular variants and outliers we minimize this sum also w.r.t. all variant selections arriving at the criterion

$$\underset{\mathbf{h}}{argmin}\,\underset{\gamma}{min} \sum_{h_i>0} -\ln f_\gamma(x_{i,h_i}). \tag{1}$$

Here, $\sum_{h_i>0} -\ln f_\gamma(x_{i,h_i})$ is the negative log-likelihood of γ for the regular variants w.r.t. the variant selection $\mathbf{h}$. The operation $\min_\gamma$ determines the m.l.e. $\gamma(\mathbf{h})$ of γ w.r.t these observations so that (1) may be rewritten as

$$\underset{\mathbf{h}}{argmin} \sum_{h_i>0} -\ln f_{\gamma(\mathbf{h})}(x_{i,h_i}).$$

Optimality of the method requires independence of all objects. Gallegos and Ritter (2006), Theorem 2.2, showed that Criterion (1) is the m.l.e. of γ and $\mathbf{h}$ w.r.t. a certain statistical model of the irregular variants that was called the *spurious-outliers model*.

Criterion (1) reduces the problem of estimating parameter and variant selection to minimizing the negative log-likelihood function of a distributional model and to a combinatorial optimization problem. Now, there are astronomically many selections, $\sum_{C\in\binom{1..n}{r}} \prod_{i\in C} b_i$; enumerating all is not feasible except for small instances and approximation algorithms are desirable. Such an algorithm is substantiated in the last-mentioned paper. Given a variant selection $\mathbf{h}$, define the negative estimated log-density

$$u_{\mathbf{h}}(i,k) = -\log f_{\gamma(\mathbf{h})}(x_{i,k}), \qquad k \in 1..b_i.$$

The basis of the algorithm is the following multi-point reduction step, a procedure that alternates parameter estimation and selection of the regular variants.

Multi-point reduction step
// Input: A selection $\mathbf{h}$;
// Output: A selection $\mathbf{h}_{\mathrm{new}}$ with improved Criterion (1)
or the response "stop."

(1) Compute the estimate $\gamma(\mathbf{h})$
(2) For each object i, determine an element $h_{new,i} \in \underset{k\in 1..b_i}{arg\,min}\, u_{\mathbf{h}(\mathbf{i},\mathbf{k})}$
(3) Determine the r objects i with minimum values $u_{\mathbf{h}}(i, h_{new,i})$ and call the corresponding selection $\mathbf{h}_{\mathrm{new}}$
(4) *If* $u_{\mathbf{h}}(i, h_{new,i}) < u_{\mathbf{h}}(i, h_i)$ for at least on i then return $\mathbf{h}_{\mathrm{new}}$; *else* "stop"

The multi-point reduction step is iterated until convergence. The variant selection obtained is self-consistent in the sense that it generates its original parameters. The optimal solution shares this property but the result of the iteration is not necessarily optimal. Therefore, the multistart method has to be applied to reduce the criterion at least to a low value.

Alternative methods for minimizing Criterion (1) are local search, the Metropolis–Hastings algorithm, the EM algorithm, and Gibbs sampling. However, to our experience the multipoint reduction step is competitive with these methods.

2.2 A Coin-Tossing Model

An interesting and important special case is a discrete model with sample space $E = (1..s)^d$ where the regular variants are generated by tossing d independent, possibly biased, s-sided coins. The parameter γ is an $s \times d$ table $\mathbf{p}$ of real numbers $p_{y,m} \geq 0$ whose columns sum to 1, the *position-specific score matrix* PSSM. Each variant $x \in E$ generates a path in this table that visits each column exactly once. Its probability is the product of the entries along the path,

$$f_{\mathbf{p}}(x) = \prod_{m=1}^{d} p_{x_m,m}.$$

Let $\mathbf{h}$ be a selection and let $n_{y,m}(\mathbf{h}) = \#\{i \,|\, x_{i,h_i,m} = y\}$ be the frequency of the outcome y at position m taken over the r selected variants of length d. These frequencies sum up to rd. The m.l.e. of the PSSM consists of the relative frequencies $n_{y,m}(\mathbf{h})/r$, $y \in 1..s$, $m \in 1..d$, and, up to a multiplicative constant, the maximum value of the likelihood function

$$f_{\mathbf{p}}(x_{\mathbf{h}}) = \prod_{i:h_i \geq 1} \prod_{m=1}^{d} p_{x_{i,h_i,m},m} = \prod_{m=1}^{d} \prod_{y \in 1..s} p_{y,m}^{n_{y,m}(\mathbf{h})},$$

cf. (1), equals their negative entropy.

The likelihood may be optimized by multistart replication of the iterative application of multi-point reduction steps. In the present context, the quantities $u_{\mathbf{h}}(i,k)$ become

$$u_{\mathbf{h}}(i,k) = -\sum_{m} \ln \frac{n_{x_{i,k,m},m}(\mathbf{h})}{r}.$$

When, in item (2) of the multi-point reduction step, a new "regular" variant is selected for object i, the relative frequencies $n_{y,m}/r$ are biased towards its current regular variant. Therefore, replacing the relative frequencies appearing in $u_{\mathbf{h}}$ with $n'_{y,m}/(r-1)$ offers a big advantage, the prime indicating omission of this object. Note that the probability estimates are now based on $r-1$ observations. Therefore, instead of the maximum likelihood, Laplace's Law of Succession should be used for estimating the probabilities $p_{y,m}$ which means that the numbers $(n'_{y,m}+1)/(r-1+s)$ replace the relative frequencies $n'_{y,m}/(r-1)$. In the extreme case of a data set consisting of one line one has the unbiased prior $1/s$.

3 Study of a Protein–Protein Interaction

GYF (glycine–tyrosine–phenylalanine) domains are highly conserved protein domains expressed in human (PERQ2), yeast (SMY2 and its paralog YPL105C), and plant (GYN4), see Kofler et al. (2005). They are characterized by two beta strands, an extended loop in between, and a successive alpha helix which is flanked by the patterns GPF (glycine–proline–phenylalanine) and GYF. A GYF domain is known to recognize proline-rich patterns in targets, the common recognition signature being PPG (proline–proline–glycine), Kofler et al. (2005).

Is it possible to detect this signature by variant analysis? More precisely: does the algorithm detect short segments in the target polypeptides which approximately match each other pairwise? The answer to this problem of *multiple local string alignment*, see, e.g., Gusfield (1997), is yes. The signature is detected in a set of protein sequences that may even be contaminated in the sense that an unknown subset of the proteins, only, act as targets. For this purpose, two data sets were compiled: Data Set A is heavily contaminated containing 100 targets and 100 non-targets, whereas data set B is moderately contaminated and contains the same 100 targets and the first 40 non-targets of Data Set A. The targets were taken from the supplemental material of Kofler et al. (2005), whereas the non-targets are the first 100 entries of the Stanford Saccharomyces Genome Database. The latter act as outliers in the data sets. We show that, despite the contamination, the interacting proteins, the common motif, and the sites where the interaction takes place can be discovered. Moreover, exact knowledge of the motif length is not necessary.

The data set consists of n polypeptides. Each polypeptide (=object) of length l amino acids gives rise to $l - d + 1$ (overlapping) segments of length d, one for each possible initial site in the sequence. These are the variants of the object and E is the d-fold Cartesian product of the set of the 20 naturally occurring amino acids. An array of r initial sites is called an *alignment*. It corresponds to a variant selection. The remaining $n - r$ sequences are outliers, i.e., considered non-targets w.r.t. the alignment. Assuming the different polypeptides to be independent as in Sect. 2.1, one may apply the foregoing theory and the modified multi-point reduction step.

Almost all pairs of residues appear very often in most medium-size and long proteins for combinatorial reasons; they are insignificant. Therefore, the smallest length d considered is three residues. The runs with $d = 3, \ldots, 6$, $r = 80, 90, \ldots, 140$, and 100,000 replications of reduction-step iterations took between one and four hours, each.

The algorithm finds two motifs: accumulations of serines and the motif PPG, see Table 2. The former are ubiquitous and prevail in the heavily contaminated data set for the motif lengths five and six and in the moderately contaminated data set for length six. The latter prevails in the moderately contaminated data set up to length five and is known to be the consensus pattern of the recognition site of the GYF domain. Since the PPG motif was identified, the same was of course true for the association sites and the target proteins of the GYF domain. Table 2 confirms the well-known fact that it is harmful to assume too many regular objects. For the length four, the result of the heavily contaminated data set A breaks down under the assumption of more regular objects than there actually are (100). By contrast, the moderately contaminated data set resists the assumption of 130 regular objects even at length five.

Table 2 Multiple sequence alignment with various motif lengths d and assumed numbers r of regular elements. *Top:* motifs found in a data set of 100 positive and 100 negative sequences (Data Set A), *bottom:* 100 positive and 40 negative sequences. Residues with specificities ≥60% at a position are shown. The lack of such a highly significant residue at some site is indicated by an *asterisk*

80	S S S S L L PPG	PPG *	S S S S S	S * * * S S
90	S S S PPG	PPG *	S * S S S	S * S * S S
100	PPG	PPG *	S * S S S	S * * * S S
110	PPG	S * S S	S * S S *	S * * * S *
120	PPG	S * S S	S * * S *	S * S * S *
130	PPG	S * S S	S * * S S	S * * * S *
140	S S S	S * S S	S * * S *	S * * * * *
d	3	4	5	6
r				
80	PPG	PPG *	* PPG *	S * S * * *
90	PPG	PPG *	* PPG *	S * S * * *
100	PPG	PPG *	* PPG *	S * * * * *
110	PPG	PPG *	* PPG *	S * * * * *
120	PPG	PPG *	* PPG *	S * * * * *
130	PPG	PPG *	* PPG *	S * * * * *
140	PPG	PPG *	S * * S *	S * * * * *

Table 3 Transposed of the PSSM for the Data Set B with parameters $r = 100$ and $d = 4$. Only residues with positive specificities at some position are shown. The consensus sequence is PPG*, the uncertain last position being mainly occupied by the hydrophobic residues isoleucine, valine, leucine, tyrosine, phenylalanine, and alanine

Res Pos	A	C	D	E	F	G	I	L	M	P	R	S	V	W	Y
0	0.00	0.00	0.00	0.00	0.00	0.00	0.00	0.00	0.00	1.00	0.00	0.00	0.00	0.00	0.00
1	0.00	0.00	0.00	0.00	0.00	0.00	0.00	0.00	0.00	1.00	0.00	0.00	0.00	0.00	0.00
2	0.00	0.00	0.00	0.00	0.00	1.00	0.00	0.00	0.00	0.00	0.00	0.00	0.00	0.00	0.00
3	0.06	0.04	0.01	0.01	0.09	0.03	0.21	0.15	0.02	0.01	0.01	0.01	0.17	0.05	0.13

The specificities of the motif PPG* are shown in Table 3. The highest specificities at the uncertain fourth position are assumed by the mostly hydrophobic side chains mentioned in the caption (an exception is alanine which is neutral).

The study shows that the method is to a certain extend robust against outliers and against an unfavorable choice of motif length and assumed number of outliers.

Acknowledgment We thank Frau Saskia Nieckau for her implementation of the algorithm.

References

Gallegos, M. T., and Ritter, G. (2006): Parameter estimation under ambiguity and contamination with the spurious model. *Journal of Multivariate Analysis 97*, 1221–1250.

Gusfield, D. (1997): *Algorithms on Strings, Trees, and Sequences; Computer Science and Computational Biology*. Cambridge University Press, Cambridge.

Kofler, M., Motzny, K., and Freund, C. (2005): GYF domain proteomics reveals interaction sites in known and novel target proteins. *Molecular and Cellular Proteomics 4*, 1797–1811. http://www.mcponline.org.

Lawrence, C. E., and Reilly, A. A. (1990): An expectation maximizatiom (EM) algorithm for the identification and characterization of common sites in unaligned biopolymer sequences. *Proteins: Structure, Function, and Genetics 7*, 41–51.

Lawrence, C., Altschul, S., Boguski, M., Liu, J., Neuwald, A., and Wootton, J. (1993): Detecting subtle sequence signals: a Gibbs sampling strategy for multiple alignment. *Science 262*, 208–214.

Liu, J. S. (1994): The collapsed Gibbs sampler in Bayesian computations with applications to a gene regulation problem. *JASA 89*, 958–966.

Liu, J. S., Neuwald, A. F., and Lawrence, C. E. (1995): Bayesian models for multiple local sequence alignment and Gibbs sampling strategies. *JASA 90*, 1156–1169.

Ritter, G. (2000: Classification and clustering of objects with variants. In W. Gaul, O. Opitz, and M. Schader, Eds., *Data Analysis, Scientific Modeling and Practical Application*, Studies in Classification, Data Analysis, and Knowledge Organization. Springer, Berlin, Heidelberg, pp. 41–50.

Ritter, G., and Gallegos, M. T.(2000): A Bayesian approach to object identification in pattern recognition. In A. Sanfeliu et al., Eds., *Proceedings of the 15th International Conference on Pattern Recognition*, vol. 2 (Barcelona), pp. 418–421.

Ritter, G., and Gallegos, M. T. (2002): Bayesian object identification: variants. *Journal of Multivariate Analysis 81*, 301–334.

Ritter, G., and Pesch, C. (2001): Polarity-free automatic classification of chromosomes. *Computational Statistics and Data Analysis 35*, 351–372.

Ritter, G., and Schreib, G. (2000): Profile and feature extraction from chromosomes. In A. Sanfeliu et al., Eds., *Proceedings of the 15th International Conference on Pattern Recognition*, vol. 2, (Barcelona), pp. 287–290.

Ritter, G., and Schreib, G. (2001): Using dominant points and variants for profile extraction from chromosomes. *Pattern Recognition 34*, 923–938.

Stormo, G. D., and Hartzell III, G. W. (1989): Identifying protein-binding sites from unaligned DNA fragments. *Proceedings of National Academy of Sciences USA 86*, 1183–1187.

Estimation for the Parameters in Geostatistics

D. Niu and T. Tarumi

Abstract In geostatistics, semivariogram is used to show spatial dependence of data on different locations. Usually the least squares method is used to estimate parameters in theoretical semivariogram, though sometimes it may not preserve the non-negative parametric properties. This paper provides a method to estimate parameters in non-negative least squares (NNLS) method for a nonlinear model. This is helpful when negative estimations are caused by a calculation error.

1 Introduction

Recently, with the progress of computer technology, software for statistical computation has been well developed in various fields. Also computers have allowed easier analysis of large data. Geographical information systems (GIS) that provide graphic visualizations of statistical analysis are helpful for reporting various kinds of information intuitively. At the same time, it is increasingly necessary to analyze with small area data for such purposes as natural disaster prevention, weather forecast, or even regional development, and public institute establishment.

In Japan, one kind of small area data so-called area mesh data. Though it has recently been utilized for its square configuration, it is not easy to directly observe each area mesh data. Sometimes we may only have the spatial data observed based on latitude and longitude. Can we create area mesh data from this kind of spatial data? To solve this problem, geostatistical principles are usually used.

Geostatistics is be used when predicting unobserved data from data on observed spatial locations. Predictions are based on the distance correlations which are indicated in semivariograms.

D. Niu(✉)
Graduate School of Natural Science and Technology, Okayama University,
1-1, Naka 3-chome, Tsushima, Okayama 700-8530, Japan, E-mail: niuniu1613@gmail.com

A. Okada et al. (eds.), *Cooperation in Classification and Data Analysis*,
Studies in Classification, Data Analysis, and Knowledge Organizations.
DOI: 10.1007/978-3-642-00668-5_13,

In this paper, we introduce geostatistics in Sect. 2 and in Sect. 3 we provide a method of estimating parameters in semivariogram model under conditions of non-negative. A numerical example is shown in Sect. 4. Conclusion and future research are described in the last section.

2 Using Geostatistics for Prediction

As a branch of spatial statistics, geostatistics is used to model the uncertainty of unknown values or to undertake spatial prediction through the generation or spatial dependence of all the data. It was essentially viewed as a means to describe spatial patterns and interpolate the value of the attribute of interest at unobserved locations when first devised by D. G. Krige. Geostatistics is much more developed in the earth sciences. It is also like time-series by converting time into locations.

2.1 Some Basic Theories in Geostatistics

Given a set $\{x_i : x_i \in D \subset \Re^d, i = 1, \ldots, n\}$ of observed data locations. Here D is a continual area in $d-$ dimensional Euclidean space $\Re^d$. Observed data value $z(x_i)$ is considered as an instance of random variable $Z(x_i)$ from random function $Z := \{Z(x); x \in D\}$.

In most cases in geostatistics, intrinsic stationarity is required. Here, we say random function Z is available with intrinsic stationarity if for any $x, x+h \in D$,

$$E[Z(x+h) - Z(x)] = 0,$$
$$\mathrm{Var}[Z(x+h) - Z(x)] = 2\gamma(h).$$

Here $2\gamma(h)$ is called variogram, and is a function of distance h, not dependent on locations, and $\gamma(h)$ is called semivariogram. If for any $x, x+h \in D$,

$$E[Z(x)] = \mu,$$
$$\mathrm{Cov}[Z(x), Z(x+h)] = C(h),$$

random function Z is said to be available with second-order stationarity. $C(h)$ is also called covariogram. The relationship between variogram and covariogram function is easy to infer as shown in the next equation

$$2\gamma(h) = 2[C(0) - C(h)].$$

In spatial process, we assume $C(h) \to 0$ when $\|h\| \to \infty$, here $\|h\|$ denotes the length of the h. It means that covariance between random functions on two points vanishes as the points furthered away from each other. A typical theoretical

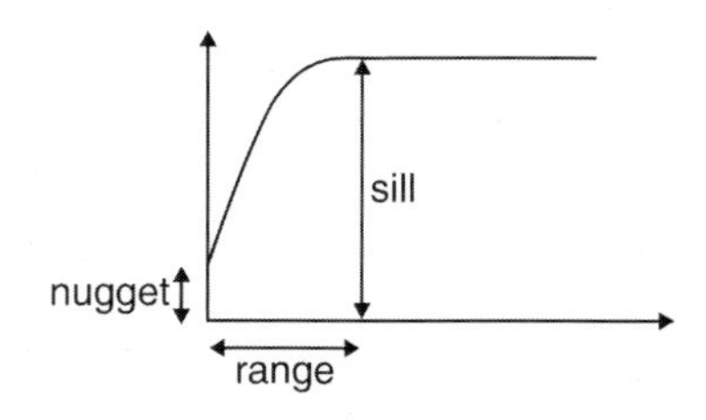

Fig. 1 Typical theoretical semivariogram model

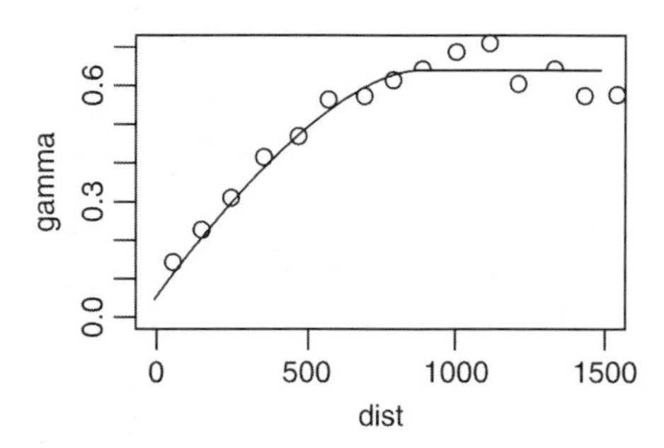

Fig. 2 Fitting theoretical semivariogram model to empirical semivariogram

semivariogram is shown in Fig. 1. Three characteristics, nugget, sill and range, are associated with semivariogram. Nugget is used to show the jump at $h \to 0^+$, also means the white noise caused by observations or data. Asymptotic value $\lim_{h\to\infty} \gamma(h)$ is called sill. And range is the minimum distance where $\gamma(h)$ reaches sill. That means if the distance of two points is over range, the covariance of the two random variables becomes zero. Here we can also know that $nugget \geq 0, sill > 0$, $range > 0$.

Generally, spatial prediction using observed data values has three steps in geostatistics.

Step 1: Calculate empirical semivariogram from observed data as (1);

$$\hat{\gamma}(h) = \frac{1}{2N(h)} \sum_{N(h)} (Z(s_i) - Z(s_j))^2, \tag{1}$$

where $N(h) = \{(i, j) : x_i - x_j = h\}$, $|N(h)|$ is the number of pairs in $N(h)$. In fact, for a given distance h, $N(h)$ must be a null set in most cases. It's an estimation lacking of stability. So instead of it we separate h into several intervals $I_1 = (0, h_1]$, $I_2 = (h_1, h_2], \ldots, I_K = (h_{K-1}, h_K]$. Let $N(h_k) = \{(i, j) : |x_i - x_j| \in I_k\}$, $k = 1, \ldots, K$.

Step 2: Fit theoretical semivariogram model to the empirical semivariogram, as shown in Fig. 2. There are some types of semivariogram models, such as spherical, exponential and Gaussian. First, people should decide which type to use at first. Then use the least squares method to solve the parameters, sill, range and nugget. That means finding θ, to minimize (2)

$$S(\theta) = \sum_{k=1}^{K} (\hat{\gamma}(h_k) - \gamma(h_k; \theta))^2, \tag{2}$$

where $\theta = (\theta_1, \theta_2, \theta_3)^T, \theta_i \geq 0, i = 1, 2, 3.$

For a general least squares calculating routing, the non-negativity of parameters, sill, range and nugget, cannot be assured. Sometimes negative estimations may be generated which cannot be used in prediction. The observed data must therefore be examined. At the same time calculation errors should be considered as well. For the former, we should think why the data lead to such kind of result. For the latter, we should change our algorithm for solving the parameters. In the next section, we provide a new idea, Non-Negative Least Squares Method (NNLS) for Nonlinear Model to solve the problem.

Step 3: Use theoretical semivariogram to predict unobserved-point data values. This method is also called kriging analysis. There are several type of kriging analyses, such as simple kriging (random function has a constant known mean and is associated with second-order stationarity), ordinary kriging (random function has an unknown constant mean and is associated with intrinsic stationarity), universal kriging (random function has a constant unknown mean within each local neighborhood). In most cases, we use ordinary kriging.

Let x_0 be a point location we want to predict. $x_1, \ldots, x_n$ are observed points around x_0. The predicted value on x_0 is denoted as a weighted linear combination:

$$Z^*(x_0) = \sum_{i=1}^{n} w_i Z(x_i), \ with \sum_{i=1}^{n} (w_i) = 1.$$

Notice that $Z^*(x_0)$ is an unbiased estimator of $Z(x_0)$ as

$$E[Z^*(x_0)] = \sum_{i=1}^{n} w_i E[Z(x_i)] = \sum_{i=1}^{n} w_i m = m = E[Z(x_0)].$$

Weight w_i, $i = 1, \ldots, n$ is solved when minimize the error variance (3)

$$\begin{aligned} \sigma^2(x_0) &= \mathrm{Var}[Z^*(x_0) - Z(x_0)] \\ &= -\gamma(x_0 - x_0) - \sum_{i=1}^{n} \sum_{j=1}^{n} w_i w_j \gamma(x_i - x_j) + 2 \sum_{i=1}^{n} w_i \gamma(x_i - x_0) \end{aligned} \tag{3}$$

under

$$\frac{\partial \sigma^2}{\partial w_i} = 0, i = 1, \ldots, n,$$

$$\sum_{i=1}^{n} w_i = 1.$$

If we put in a Lagrange parameter λ, w_i can be solved from next normal equations (4),

$$\begin{pmatrix} \gamma_{11} & \dots & \gamma_{1n} & 1 \\ \vdots & \ddots & \vdots & \vdots \\ \gamma_{n1} & \dots & \gamma_{nn} & 1 \\ 1 & \dots & 1 & 0 \end{pmatrix} \begin{pmatrix} w_1 \\ \vdots \\ w_n \\ \lambda \end{pmatrix} = \begin{pmatrix} \gamma_0 1 \\ \vdots \\ \gamma_0 n \\ 1 \end{pmatrix}. \quad (4)$$

Here $\gamma_{ij} = \gamma(x_i - x_j)$, $\gamma_{0i} = \gamma(x_i - x_0)$, $i, j = 1, \dots, n$. In the normal equations system, the left side provides the dissimilarity among the observed locations, and the right one shows that between the observed locations and predicting locations. Sometimes we use LU decomposition in the algorithm solving this problem.

Above three steps are mainly used in prediction problems. In the next section, for the Step 2, we provide a method for estimating theoretical semivariogram model according to the empirical semivariogram.

3 Non-negative Least Squares Method for Nonlinear Model (NNLS)

3.1 Idea of NNLS for Nonlinear Model

As we have said, sill, range and nugget are non-negative in geostatistics. To estimate these parameters is a problem of constrained least squares. Here the constrained condition is non-negative. The parameters can be set as follows:

$$\theta_i = \varphi_i^2, \quad i = 1, 2, 3.$$

Let

$$\phi = (\varphi_1, \varphi_2, \varphi_3)^T, \quad \theta = (\theta_1, \theta_2, \theta_3)^T.$$

Here $\theta_1, \theta_2, \theta_3$ indicate nugget, sill and range respectively. Then, the squares sum $S(\theta)$ in (2) is become (5)

$$G(\phi) = \sum_{k=1}^{K} [\hat{\gamma}_i - \gamma(h_k, \phi)]^2 = [\hat{\gamma} - \gamma(h, \phi)]^T [\hat{\gamma} - \gamma(h, \phi)]. \quad (5)$$

Here, $\hat{\gamma}_i$ is empirical semivariogram. γ is theoretical semivariogram with parameters ϕ. Using Newton–Raphson method to solve the parameter in $S(\theta)$, we have

$$\frac{\partial S}{\partial \theta_i}|_{\theta=\theta^*} = 0$$

the same as

$$\frac{\partial G}{\partial \varphi_i}|_{\phi=\phi^*} = 0.$$

With the kth approximate value $\theta^{(k)}$, we can get $\Delta\theta$ from (6)

$$\nabla^2 S(\theta^{(k)}) \cdot \Delta\theta = -\nabla S(\theta^{(k)}). \tag{6}$$

And the $(k+1)$th approximate value is

$$\theta^{(k+1)} = \theta^{(k)} + \Delta\theta.$$

Using Gauss–Newton method, the equations are

$$\begin{aligned} \mathbf{U}^{\mathbf{T}}\mathbf{U}\Delta\phi &= \mathbf{U}^{\mathbf{T}}[\hat{\gamma} - \gamma(\mathbf{h}, \phi)], \\ \phi^{(k+1)} &= \phi^{(k)} + \alpha\Delta\phi, \end{aligned} \tag{7}$$

where α is step-size. For the respective stability, a step-size $\alpha(0 < \alpha \leq 1)$ is multiplied to the incremental vector $\Delta\phi$. α is used to assure S to drop off. U is the derivative matrix shown as

$$U = \begin{pmatrix} \frac{\partial g_1(\phi)}{\partial \varphi_1} & \frac{\partial g_1(\phi)}{\partial \varphi_2} & \frac{\partial g_1(\phi)}{\partial \varphi_3} \\ \vdots & \vdots & \vdots \\ \frac{\partial g_K(\phi)}{\partial \varphi_1} & \frac{\partial\partial g_K(\phi)}{\partial \varphi_2} & \frac{\partial g_K(\phi)}{\partial \varphi_3} \end{pmatrix} = \begin{pmatrix} \frac{\partial \gamma_1(\theta)}{\partial \theta_1} \cdot 2\varphi_1 & \frac{\partial \gamma_1(\theta)}{\partial \theta_2} \cdot 2\varphi_2 & \frac{\partial \gamma_1(\theta)}{\partial \theta_3} \cdot 2\varphi_3 \\ \vdots & \vdots & \vdots \\ \frac{\partial \gamma_K(\theta)}{\partial \theta_1} \cdot 2\varphi_1 & \frac{\partial \gamma_K(\theta)}{\partial \theta_2} \cdot 2\varphi_2 & \frac{\partial \gamma_K(\theta)}{\partial \theta_3} \cdot 2\varphi_3 \end{pmatrix}$$

$$= A \cdot \begin{pmatrix} 2\varphi_1 & 0 & 0 \\ 0 & 2\varphi_2 & 0 \\ 0 & 0 & 2\varphi_3 \end{pmatrix} = A \cdot \mathrm{diag}(2\varphi_1, 2\varphi_2, 2\varphi_3)$$

and here A is the derivative matrix of γ to θ shown as

$$A = \begin{pmatrix} \frac{\partial \gamma_1(\theta)}{\partial \theta_1} & \frac{\partial \gamma_1(\theta)}{\partial \theta_2} & \frac{\partial \gamma_1(\theta)}{\partial \theta_3} \\ \vdots & \vdots & \vdots \\ \frac{\partial \gamma_K(\theta)}{\partial \theta_1} & \frac{\partial \gamma_K(\theta)}{\partial \theta_2} & \frac{\partial \gamma_K(\theta)}{\partial \theta_3} \end{pmatrix}.$$

We get ϕ directly here, and its square is θ. Notice that parameters θ_i or φ_i should not be 0 or the left side in (7) must be singular. As a supplementary, we discuss the case of the parameters equal 0 in our algorithm. In the recurrence progress, set the lower limitation as ϵ.

If some parameter such as $|\varphi_i| < \epsilon$, begin:

(a) At the case φ_i is allowed to be 0:

First, let $\varphi_i = 0$ and use NNLS on other parameters. The result is set S_1 here. Second, let $\varphi_i = \epsilon \times \varphi_i / |\varphi_i|$, then use NNLS, we can get a result S_2. Comparing the result S_1 and S_2, select values in the smaller one as the estimation of parameters.

(b) At the case φ_i is not allowed to be 0:

Calculate only the second step and obtain the estimation of parameters.

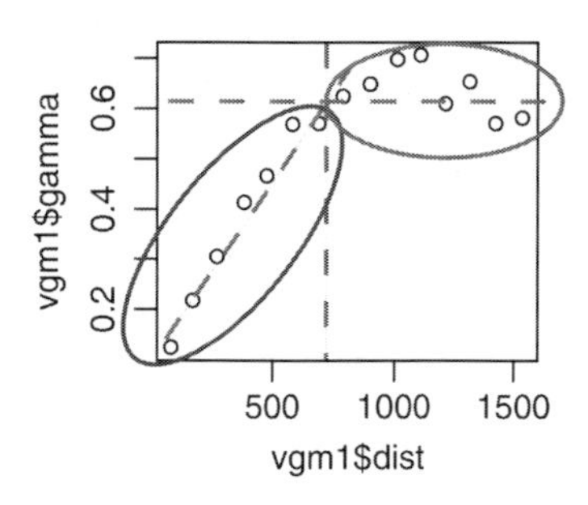

Fig. 3 Image of calculating the initial values

Table 1 Artificial empirical semivariogram with 15 points

Distance	1	2	3	4	5	6	7	8
Gamma	1.49199	2.29953	3.57124	4.96044	7.33973	6.04643	8.43456	8.96850
Distance	9	10	11	12	13	14	15	
Gamma	10.59282	10.39316	11.97131	11.31546	11.84075	10.47581	9.32058	

Our algorithm is based on Gauss–Newton method. The convergence condition is the same with Gauss–Newton method.

3.2 Decision of Initial Values

When using this algorithm of least squares method, in the beginning we shall decide the initial values of parameters. Likely initial values of parameters may lead to true parameters more efficiently. Usually the initial values are decided by user. It can also be calculated as in Fig. 3: The empirical semivariogram is obtained as points in Fig. 3. We divide the empirical semivariogram points into two sets by the midpoint of distance. Using regression analysis on the left sets, and a regression line can be obtained. The intercept is set as the initial value of nugget. If the intercept is solved as a negative value, initial of nugget is set to be 0. We calculate the average of the second sets, and set it as the initial of sill. The horizontal axis of the intersection point of the lines is set as the initial value of range. This is a simple idea but works well in most cases.

4 Numerical Example

In this section we offer a numerical example of using our algorithm in fitting theoretical semivariogram. The artificial empirical semivariogram is shown in Table 1.

When using the ordinal least squares method, the result of parameters are shown in the left of Table 2. As the nugget parameter is a negative value, these parameters

Table 2 Result of using least squares method

	NLS	NNLS
Sill	11.2571251	11.25749
Range	12.0451452	12.04242
Nugget	−0.2999618	1.966407e-07
Error	8.812072	10.17104

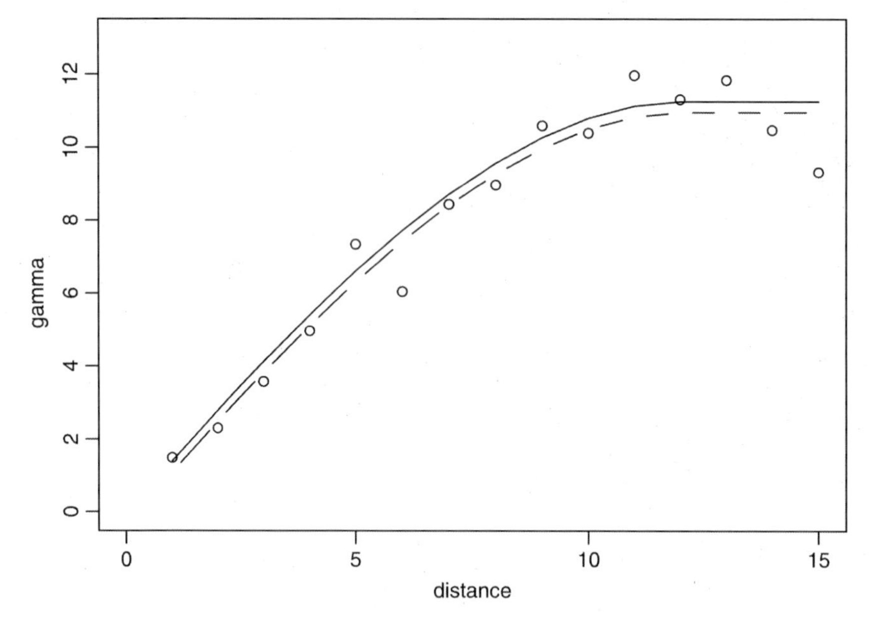

Fig. 4 Result of non-negative least squares and nonlinear least squares method (*circles* empirical semivariogram points, *solid line* NNLS, *dashed line* NLS)

can't be used in kriging analysis for predicting values on unobserved places. The results of using our algorithm are shown in the right-hand of Table 2. Of course the error becomes a little larger. But we can use it to predict data on other locations (Fig. 4).

5 Conclusion and Future Research

The semivariogram is used to show spatial dependence. When some parameters of semivariogram are negative in value, we should first examine the research data. Our algorithm is used in cases occurred due to calculation errors. The parameters are then used to do kriging analysis. This algorithm is used in our system for predicting area mesh data. For improving the accuracy of prediction, we will check the influence analysis resulting on geostatistics, and insert it into the area mesh statistics.

Identifying Patients at Risk: Mining Dialysis Treatment Data

T. Knorr, L. Schmidt-Thieme, and C. Johner

Abstract A growing number of patients suffer from end stage renal failure. For these patients either transplantation or regular dialysis treatment is necessary for survival. But the cost of providing dialysis care is high and the survival span of patients on dialysis is significantly lower than for patients with transplants. During dialysis two dozen parameters can be observed on a regular basis. Analyzing these data with data mining algorithms can help to identify critical factors for patient survival.

The paper provides a brief review of state-of-the-art methods for predicting patient risk as well as some new ideas. Its main contribution lies in the incorporation of the underlying temporal structure of the dialysis data where other studies consider only aggregated values. All methods are evaluated on real-world data from dialysis clinics in Southern and Eastern Europe.

1 Introduction

A growing number of patients suffer from end stage renal failure. For these patients either transplantation or regular dialysis treatment is necessary for survival. But the cost of providing dialysis care is high, and the survival span of patients on dialysis is significantly lower than for patients with transplants. During dialysis two dozen parameters can be observed on a regular basis. Analyzing these data with data mining algorithms can help to identify critical factors for patient survival.

The median survival span of dialysis patients in Europe is only around 5 years (QuaSi-Niere 2005). Given the long waiting period of more than 3 years for transplants, the target of our study was to use the available data to gain insights about the

T. Knorr(✉)
Institut für Informatik, Universität Freiburg, 79110 Freiburg, Germany; and Calcucare GmbH, Kaiser-Joseph-Str. 274, 79098 Freiburg, Germany, E-mail: till.knorr@googlemail.com

A. Okada et al. (eds.), *Cooperation in Classification and Data Analysis*, Studies in Classification, Data Analysis, and Knowledge Organizations.
DOI: 10.1007/978-3-642-00668-5_14,

rest lifespan of the patients, e.g., how many years a patient can expect to live after entering dialysis treatment.

Discretizing the lifespan into two classes with respect to above or below median leaves us with the supervised learning task of classification: predict the class of an unlabeled instance (a patient), given its features (the dialysis treatment data) and a training set of labeled instances. Accurate predictions could be used, for example, to prefer the patients at risk when a new donor kidney becomes available.

All our experiments have been evaluated on an industrial data-set donated by Calcucare GmbH, Freiburg. To the best of our knowledge this is the largest dialysis treatment data-set in the literature consisting of over 100,000 dialysis sessions from 747 patients.

The rest of the paper is organized as follows. Section 2 gives a more detailed view of the medical background and discusses related work. Section 3 showcases the data-set we used, and explains the various measurements that are recorded. Sections 4–6 introduce classification models, first a standard attribute-value learner, then an enhanced model that uses the temporal structure of the data and finally a combination of both. Section 7 concludes the paper with a summary and an outlook to possible future research.

2 Background and Related Work

2.1 Kidney Failure and Dialysis Treatment

The human kidney is responsible for a variety of crucial tasks of the human body, such as metabolic processes and the cleaning of toxins from the blood. Several factors are known for damaging the kidneys, e.g., drug or alcohol abuse, diabetes, high blood pressure.

When the kidneys can no longer filter metabolic toxins from the blood and eliminate them through the urine this condition is known as end stage renal disease. In this stadium there are two possibilities to keep the patient alive. The first option is that the patient receives a surgery to implant a healthy donor kidney. However, donor organs are not available in abundance and other medical reasons may make a transplantation impossible.

The other possibility is to receive repeated dialysis treatment where the dialysis process replaces the kidneys' functions. During hemodialysis the patient's blood is passed through an extra-corporal circuit connected to the dialyzer machine. The dialyzer then filters the urea from the blood by means of a semi-permeable membrane. The cleaned blood is returned to the patient's body. This treatment takes approximately 4 h and is usually performed at a dialysis clinic three to four times a week.

When entering the dialysis treatment for the first time a patient in Germany has a median rest lifespan of around 5 years. While this is partly due to the fact that these patients often suffer from various other diseases as diabetes and heart problems, the survival span is still significantly lower than that of patients with a renal transplant. This indicates that there exist ways to improve the quality of the dialysis.

2.2 Related Work

Kusiak et al. (2005) were among the first to study data mining in the field of dialysis treatment. The parameters they observed in clinics from the USA. are similar to ours. While Kusiak et al. did not consider the temporal structure of the data, they point this out as possible further research. Their main contribution was to show the effectiveness of data mining methods in this domain. Using a modified rule learner, they were able to extract rules that made sense to a medical expert, and achieved an overall accuracy between 75% and 85%.

The study by Bellazzi et al. (2005) has a different problem setting. Their goal was not to predict the long term outcome for a patient, e.g., the survival span, but whether a given dialysis session will result in a failure. A failed sessions occurs for example when the blood pressure of the patient cannot be controlled sufficiently. In their study they also consider temporal aspects of the recorded data. Their model can find rules of the form "when blood pressure is low while heart rate is high then complication".

3 The Data-set

During dialysis a large number of parameters can be monitored on a regular basis, e.g., blood pressure, heart rate. The data-set of this study was extracted from databases of Calcucare GmbH, Freiburg, and consists of data from more than 1,000 patients from clinics operated by Fresenius AG. For each of the patients, values of at least 100 sessions have been recorded with a total of more than 100,000 sessions. Other studies report 188 patients (Kusiak et al. 2005) and 33 patients (Bellazzi et al. 2005). Table 1 on page 134 shows an outline of the data which can be differentiated into two categories:

- *One-time-measurements:* parameters that are usually assessed only once, e.g., patient age, sex, days on dialysis treatment. Rarely measured chemical lab values (e.g., albumin) are added as average values.
- *Regular measurements:* values monitored on every dialysis session, stored as average value of the 4-h session period.

3.1 Target Class Label

We adopt the binary class from Kusiak et al. (2005) as prediction target: whether or not the patient survived for more than the median number of days on dialysis

Table 1 Example of available parameters

Patient profile and rare measurements							
Pat ID	Age	Sex	Height	Comorbid	Days	Albumin	...
0001	53	M	175	Diabetes	1,100	30 ($\mathrm{g\,l^{-1}}$)	...
Regular measurements							
Blood pressure		Heart rate	Blood vol.	Weight loss	Blood flow		...
130:70 (mmHG)		80 ($\mathrm{s^{-1}}$)	92 (l)	1.95 (kg)	390 ($\mathrm{ml\,min^{-1}}$)		...

treatment. In our case this was 1,800 days and the classification is calculated as follows:

- If $survivalspan \geq 1{,}800 \Rightarrow class = high$.
- If $survivalspan < 1{,}800 \wedge deceased \Rightarrow class = low$.
- If the patient was less than 1,800 days on treatment and is still alive, he or she could become either class and is excluded from the study.

With the exclusions according to rule 3 the resulting distribution was 53% high and 47% low with a total of 747 patients.

4 Standard Model

For the first experiment we used aggregated values as features to train a classifier, similar to Kusiak et al. (2005). In particular we calculated minimum, maximum and median values for all of the around 30 session variables for the first 100 dialysis sessions of each patient. Adding the one-time-measurements resulted in an approx. 40 dimensional feature vector (see Table 2, page 135).
We trained three different types of classifiers on our data-set, namely:

- A Support Vector Machine (RBF Kernel, $\gamma = 0.2$)
- The rule learner "The Ripper" (4 folds, 10 optimizations)
- And the meta classifier AdaBoost (Base classifier: Decision Stump)

We used Weka (Witten and Frank 2005) implementations of these classifiers. Experiments were done with ten times 10-fold cross-validation, e.g., we repeated a 10× cross validation ten times and then reported the average. The results are shown in Table 3 on page 135.

As Table 3 shows, adding features to the training data improved classification accuracy. Base features like age and sex alone could do well enough with around 70% accuracy. The features from the session data also were able to predict at that level, but the best results could be achieved by combining the two. While JRip performed worst on all data-sets we still included it since the results of the rule learner are understandable by humans and the rules extracted were confirmed by our medical expert to be known facts (e.g., benefit of high albumin level).

Table 2 Example patient feature vectors

No.	Class	Age	Sex	Albumin	Median heart rate	Max weight loss	...
1	High	70	F	31	65	2.5	...
2	Low	55	M	25	75	1.7	...

Table 3 Results of the standard model, separated by data-set and classifier

Used parameters	Classifier accuracy in %		
	SVM	AdaBoost	Ripper
Only median session values	64.0	62.5	59.5
Only base values	67.7	69.8	65.1
Min, median, max session values	70.0	67.6	63.2
Base and median session values	70.5	72.7	64.3
All available values	73.1	75.8	69.1

5 Temporal Model

Instead of averaging over all the treatment results we wanted to extract features from the dialysis time series directly. To give an intuition why the shape of the whole series might correlate with the survival span, let us have a look at an example: The time series in Fig. 1 on page 136 show the blood flow over 100 dialysis sessions from four patients.

It is obvious to humans that the time series of the high patients are smoother while the time series of the class low patients seem to be more unsteady. This similarity could be exploited by, for example, a nearest-neighbour classifier. First we need to define the notions of time series and similarity.

A *time series* of length n is an ordered set of n real valued numbers:

$$r_1, \ldots, r_n \quad r_i \in \mathbf{R} \quad \forall i = 1 \ldots n$$

Every patient has around 30 time series of length 100, one for every session parameter. One can compute the similarity of two time series $Q = q_1, \ldots, q_n$ and $C = c_1, \ldots, c_n$ by means of a measure such as *Euclidian distance:*

$$D(Q, C) = \sqrt{\sum_{i=1}^{n} (q_i - c_i)^2}.$$

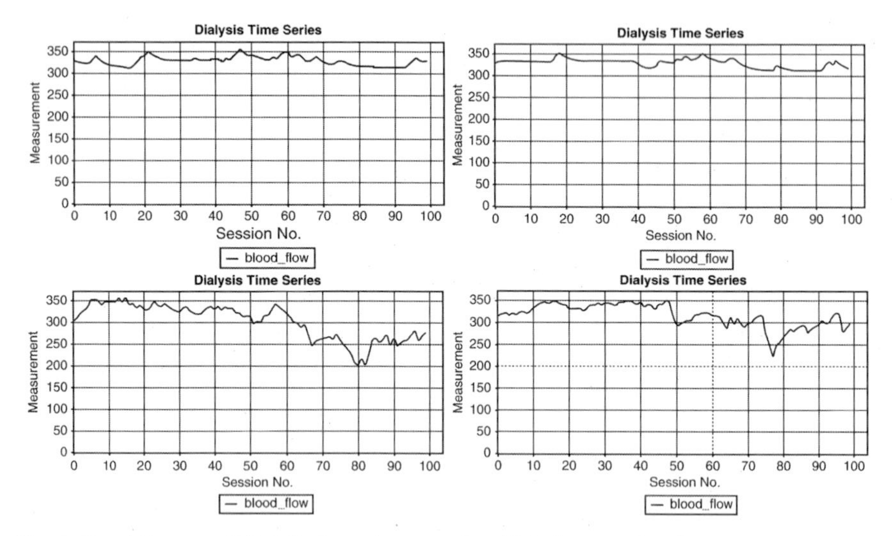

Fig. 1 Blood flow over 100 dialysis session, *top:* class high, *bottom:* class low

Table 4 Results kNN with euclidian distance, $k = 7$

Time series	kNN performance in %		
	Accuracy	Precision (class low)	Precision (class high)
Blood flow	64.3	69.1	62.3
Difference syst. bp after 1 h	64.1	56.5	72.6
Difference dias. bp after 1 h	55.7	49.9	72.0

5.1 Nearest Neighbour Classifier

The idea was to train a k-Nearest-Neighbour classifier for every time series by using an appropriate similarity measure. The kNN classifier will output the most likely class of a patient given its neighbors with respect to the similarity measure. While the kNN already predicts a class for every patient this prediction is solely based on one time series. By combining predictions of different kNN classifiers into a meta classifier, e.g., by simple voting, the whole available data can be used. Such a mixed classifier can be expected to outperform the single base classifiers. Kadous and Sammut (2005) use this kind of meta-features for multivariate time series classification. We will show the results of the single kNN classifiers first, before we proceed with the mixed model in Sect. 6.

The most predictive time series were the blood flow and the drop in systolic and diastolic blood pressure 1 h after the start of the dialysis session. The kNN classifiers were evaluated by 10 times 10× cross-validation.

As Table 4 on page 136 shows, the performance of the euclidian kNN classifier was rather low on all time series. This might be a problem of the similarity measure in use. It has been known that flexibility in the time axis often brings benefits over

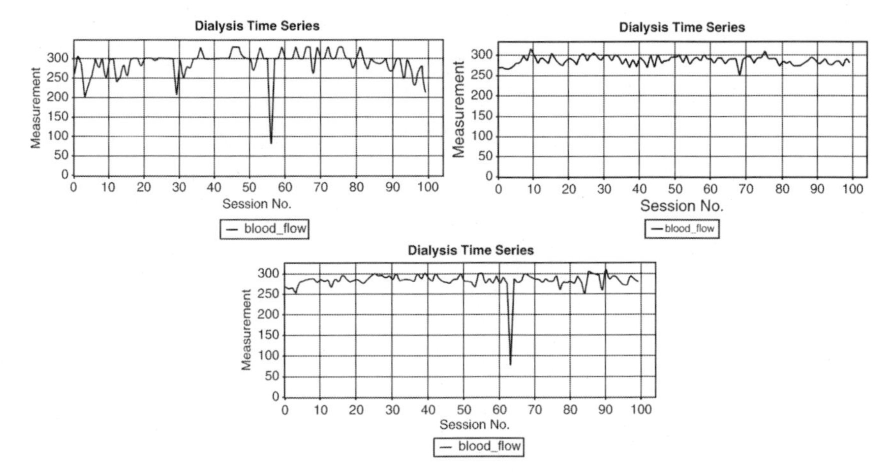

Fig. 2 *Left:* query time series, *middle:* 1-NN euclidian, *right:* 1-NN DTW

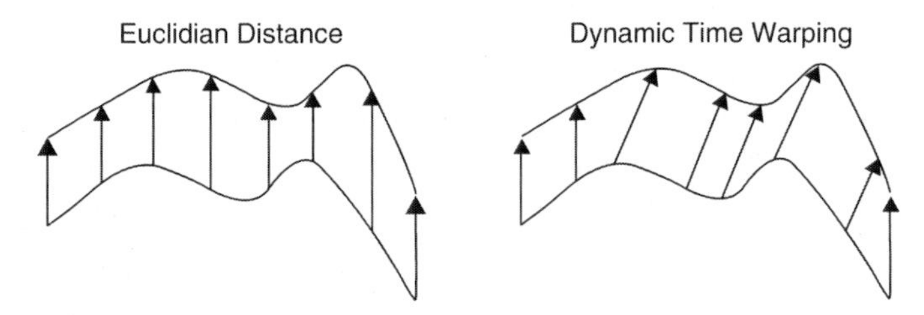

Fig. 3 Comparison euclidian distance and DTW

simple euclidian distance (Keogh and Kasetty 2002). In our case, whether a sudden drop in blood flow occurs in session number 10 or 23 is not very important, the similarity should be measured in terms of "if" and "how often" such a drop occurs. This intuition is illustrated in Fig. 2.

Figure 2 shows the blood flow time series of three patients. The series on the left is from a class low patient. In the middle is the most similar blood flow according to the euclidian metric which was a high patient and therefore misclassified by a 1-NN. On the right side is the most similar patient returned by the DTW (dynamic time warping) measure which we will introduce in the next section. This is a class low patient and thus a correct classification.

5.2 Dynamic Time Warping

The DTW similarity allows it to compare two values that did not occur at the same time point. See Fig. 3 for an example.

Given two time series $Q = q_1, \dots, q_n$ and $C = c_1, \dots, c_m$ of length n respectively m, DTW(Q,C) can be calculated by the following algorithm:

Table 5 Results kNN with DTW distance

Time series	kNN accuracy in %
Blood flow	67.9
Difference syst. bp after 1 h	73.2
Difference dias. bp after 1 h	71.5

- Initialize a matrix D with euclidian distances between every pair of points:

$$D = D[i][j], \quad i = 1, \dots n, \quad j = 1, \dots m, \qquad D[i][j] = \sqrt{(q_i - c_j)^2}.$$

- Search a way from the top right corner of the matrix$(D[n][m])$ to the bottom left corner $(D[0][0])$ by applying the steps *down* $(D[i-1][j])$, *left* $(D[i][j-1])$ or *down-left* $(D[i-1][j-1])$. This path has to be a minimum cost path with respect to the cost function

$$\text{cost(path)} = \sum_{D[i][j] \text{ on path}} D[i][j].$$

- Finding the minimum cost path can be done recursively:

$$\gamma(i, j) = D[i][j] + \min\{\gamma(i-1, j), \gamma(i, j-1), \gamma(i-1, j-1)\}.$$

- The final DTW distance is $DTW(Q, C) = \gamma(n, m)$.

DTW can be shown to be in $o(n \cdot m)$ and since it does not obey the triangular inequation the calculation has to be repeated for every pair of patients. Still, the whole process of finding the nearest neighbours can be sped up by using a lower-bounding function called LB_Keogh that can be computed in linear time (Keogh 2002). The true DTW distance for a candidate has to be computed only if the LB distance of that candidate is below the DTW distance of the current nearest neighbour.

We implemented DTW and LB_Keogh in Java and used them to train kNN classifiers similar to the ones using euclidian distance. Table 5 shows the results.

As expected using DTW instead of euclidian distance provided overall improvements on classification accuracy. In the next section we will see if a mixed model can perform even better.

6 Mixed Model

For the final experiment we used kNN classifiers with the DTW similarity measure as preprocessing step to extract binary features from each time series. These were combined with the base features like age, lab values, etc. The setup was the same as in Sect. 4. Table 6 compares the results.

All three classifiers improved accuracy when using the kNN features instead of aggregated data in the case where base data is not included. When adding base data,

Table 6 Results of the mixed model experiment

Included features	Classifier accuracy in %		
	SVM	AdaBoost	JRip
Only aggregated session data	70.0	67.6	63.2
Only kNN predictions	73.1	71.1	67.3
Base + aggregate	73.1	75.8	69.1
Base + kNN predictions	73.7	76.4	70.4

however, the difference between the feature sets became insignificant. Furthermore, only 6.3% of the around 25% misclassifications fall in a "don't care" interval in terms of total days on dialysis treatment, e.g., $1{,}800 \pm 100$ days.

7 Summary and Outlook

This work considered data mining in the medical setting of hemodialysis treatment. We gave a brief review of previous approaches in this domain, like Kusiak et al. (2005) and Bellazzi et al. (2005) which introduced the problem settings of survival span and dialysis failure prediction. The main disadvantage of these studies is the small size of the data-sets. Luckily we did have more data available, which enabled us to recreate some of the experiments and to report results that should be more reliable.

We then went on to incorporate a temporal analysis of the data into the standard model which was suggested by Kusiak et al. (2005) to be a shortcoming of their model. Our temporal model included predictions by nearest neighbour classifiers measuring the similarity of time series with dynamic time warping distance. These predictions were then joined in a larger mixed model to achieve the final outcome. The temporal model turned out to be not significantly better than the standard model, but many extensions are possible when considering time series data.

For example, instead of looking at whole series similarity, one could investigate the local structure of a time series. By extracting subsequences with a sliding window, reoccurring motifs in the time series can be found (Chiu et al. 2003). An advantage over the kNN features is, that these motifs might be understandable by a human expert and provide further insight into questions like why a motif indicates an above or below median survival-span.

References

BELLAZZI, R., LARIZZA, C., MAGNI, P., and BELLAZZI, R. (2005): Temporal data mining for the quality assessment of hemodialysis services. *Artificial Intelligence in Medicine* 34(1): 25–39.

CHIU, B., KEOGH, E., and LONARDI, S. (2003): Probabilistic discovery of time series motifs. In *The 9th ACM SIGKDD International Conference on Knowledge Discovery and Data Mining.* Washington, DC, USA. pp. 493–498.

KADOUS, M. and SAMMUT, C. (2005): Classification of multivariate time series and structured data using constructive induction. *Machine Learning* 58:179–216.

KEOGH, E. (2002): Exact indexing of dynamic time warping. In *Proceedings of the 28th VLDB Conference,* Hong Kong, China.

KEOGH, E. and KASETTY, S. (2002): On the need for time series data mining benchmarks: a survey and empirical demonstration. In D. Hand, D. Keim, and R. Ng, editors, *Proceedings of the 8th ACM SIGKDD International Conference on Knowledge Discovery and Data Mining*, pp. 102–111. ACM Press, 2002.

KUSIAK, A., DIXON, B., and SHAH, S. (2005): Predicting survival time for kidney dialysis patients: a data mining approach. *Computers in Biology and Medicine,* 35(4):311–327.

QuaSi-Niere: Annual Report on Dialysis Treatment and Renal Transplantation in Germany for 2004/2005, Website: http://www.quasi-niere.de.

WITTEN, I.H. and FRANK, E. (2005): *Data mining: practical machine learning tools and techniques*, 2nd Edition. Morgan Kaufmann, San Francisco.

Sequential Multiple Comparison Procedure for Finding a Changing Point in Dose Finding Test

H. Douke and T. Nakamura

Abstract Changing point analysis in many dose finding tests has been applied to examine the change of a response with increasing dose levels of an experimental compound. In this study, we assume that the group observations at each dose level are sequentially obtain and a response shows a linear tendency at an early stage with increasing dose levels. Then we propose a sequential multiple comparison procedure to identify the minimum effective dose level to first separate from a linear response by sequentially comparing each dose level with the zero dose level. Furthermore, we present a formulation to determine a critical value, the power of the test and a procedure to determine the necessary sample size in our sequential test.

1 Introduction

In engineering, the dose finding test have frequently been used to examine how a response changes with increasing dose levels of an experimental compound. Chen and Gupta (2000) discussed the changing point analysis based on max-t statistic using the observations obtained for all dose levels. In this study, we assume that the group observations at each dose level are sequentially obtained and the response has a linear tendency at an early stage. Then we devise a multiple comparison procedures to identify the minimum effective dose (MED) level in order to find a changing point in which the response separates from the tendency at some level with increasing dose levels.

Representative multiple comparison procedures have presented by Tukey (1953), Dunnett (1955), Scheffé (1953), and other, as methods to find simultaneously all the differences in efficacy among several dose levels. Bartholomew (1959a,b), Williams (1971, 1972) discussed multiple comparison procedures for the dose finding test

T. Nakamura(✉)
Tokai University, Hiratsuka, Kanagawa 259-1292, Japan,
E-mail: tomonaka@keyaki.cc.u-tokai.ac.jp

A. Okada et al. (eds.), *Cooperation in Classification and Data Analysis*,
Studies in Classification, Data Analysis, and Knowledge Organizations.
DOI: 10.1007/978-3-642-00668-5_15,

under the assumption in which the change of a response is monotone. From a viewpoint of the power of the test, Williams (1971, 1972), Dunnett and Tamhane (1991), Marcus et al. (1976) have proposed the step-down procedures, i.e., the methods to find all the differences among the population means with reducing sequentially the number of hypotheses from the family of hypotheses. Recently, Dunnett and Tamhane (1992, 1995) developed the step-up procedure that identifies the MED level by successively comparing each ordered dose levels with a control level of the response.

In our multiple comparison procedure, we first set up null hypothesis at each step under an assumption in which the series of the population means on the response has a linear tendency (monotone increasing or monotone decreasing) with increasing dose levels. Then, we devise a stepwise test to identify the MED level in which the response first separates from the linear tendency at some level by sequentially comparing each dose level with the zero dose level. Thus we set up alternative hypothesis and statistic to identify the MED level at each step in the sequential test. If we can find the MED level at an early stage in the sequential test, it is possible to terminate the procedure in the dose finding test after a few group observations up to the dose level. When it is difficult or expensive to collect the observations, the multiple comparison procedure is effective from an economical point of view.

In the procedure, we present how to determine a critical value for the two-sided step wise test under the null hypothesis with a predefined type I familywise error rate. Furthermore, we formulate the power of the test and conduct how to determine the necessary sample size at each dose level for satisfying the specified power of the test. In practice, we give a simulation study to compare the powers of test and the necessary sample sizes on the various configurations of population means. Finally, we apply our sequential procedure to a case study on a concrete technology in order to find a suitable amount of admixture in efficacy on the consistency of mortar in the workability.

2 Sequential Multiple Comparison Procedure

We will consider the two-sided and one-sided step wise test to sequentially find the separation from a linear response with increasing dose levels.

First we specify the number of dose levels $K+1$ and a type I familywise error rate α. Suppose a set of increasing dose levels by $0, 1, \ldots, K$, where 0 corresponds to a zero dose level. Assume that n group observations at each dose level are sequentially obtained, and call "ith stage" the ith$(i=0,\ldots,K)$ dose level to sequentially test. Suppose that all observations $x_{ij}(i = 0,\ldots,K,\ j = 1,\ldots,n)$ are mutually independent and the response $X_i(i = 0,\ldots,K)$ at the ith stage is distributed according to normal distribution with a population mean $\mu_i(i = 0,\ldots,K)$ and an unknown variance σ^2. We will write the distribution as

$$X_i \sim N(\mu_i, \sigma^2),\ \ i = 0,\ldots,K. \tag{1}$$

The MED is defined as MED $= \min\{i : \mu_i \neq \mu_0 + ia\}$. In this study, we call "step" to test sequentially for the following null hypotheses $H_0^{(i)} (i = 1, \ldots, K)$ against the alternative hypotheses $H_1^{(i)} (i = 1, \ldots, K)$ as

$$i\text{th step} \quad H_0^{(i)} : \mu_i = \mu_0 + ia (= \mu_{i-1} + a), \quad H_1^{(i)} : \mu_i \neq \mu_0 + ia, \quad (2)$$

where a is a constant concerned with increasing or decreasing tendency. If $H_0^{(1)}, \ldots, H_0^{(i)}$ are retained up to the ith step in our sequential test, we can let $\mu_0 = 0,\ \mu_1 = a,\ \mu_2 = 2a, \ldots,\ \mu_i = ia$ without loss of generality. The sample mean $\bar{x}_i$ at the ith stage is $\bar{x}_i = \sum_{j=1}^{n} x_{ij}/n \sim N(\mu_i, \sigma^2/n)(i = 0, \ldots, K)$. Under the hypothesis $H_0^{(i)}$ at the ith step, $\bar{x}_i - (\bar{x}_0 + ia) \sim N(0, 2\sigma^2/n)$. Thus $Y_i = \{\bar{x}_i - (\bar{x}_0 + ia)\}/\sqrt{2\sigma^2/n} \sim N(0,1)(i = 1, \ldots, K)$ and $U_i = \sum_{j=1}^{n} (x_{ij} - \bar{x}_i)^2/\sigma^2 \sim \chi^2(n-1)(i = 0, \ldots, K)$ where $\chi^2(n-1)$ is the χ^2 distribution with $(n-1)$ degrees of freedom and express the probability density function of U_i as $g(u_i)$. Then $U_0 + U_i \sim \chi^2[2(n-1)]$. Thus, T_i at the ith step is denoted by

$$T_i = \frac{Y_i}{\sqrt{\frac{U_0+U_i}{2(n-1)}}} \sim t[2(n-1)], \quad i = 1, \ldots, K, \quad (3)$$

where $t(2(n-1))$ is the t distribution with $2(n-1)$ degrees of freedom.

When a critical value r and the t-values $t_1, \ldots, t_K$ of $T_1, \ldots, T_K$ are given, the decision rules are as follows:

- ith step$(i = 1, \ldots, K-1)$

  (a) If $|t_i| \leq r$, then one retains the decision and tests again based on $|t_{i+1}|$.
  (b) If $|t_i| > r$, then one rejects the hypothesis $H_0^{(i)}$ and terminates the test.

- Kth step

  (c) If $|t_K| \leq r$, then one accepts the hypothesis $H_0^{(K)}$.
  (d) If $|t_K| > r$, then one rejects the hypothesis $H_0^{(K)}$.

Under $H_0^{(1)}, \ldots, H_0^{(K)}$, we define a type I familywise error rate α as

$$\begin{aligned}\alpha = {} & \Pr(|T_1| > r) + \Pr(|T_1 \leq r, |T_2| > r) + \Pr(|T_1| \leq r, |T_2| \leq r, |T_3| > r) \\ & + \cdots + \Pr(|T_1| \leq r, \ldots, |T_{K-1}| \leq r, |T_K| > r). \end{aligned} \quad (4)$$

We can express

$$1 - \alpha = \Pr(|T_1| \leq r, \ldots, |T_K| \leq r). \quad (5)$$

3 Critical Value

We would like to decide a critical value r so as to satisfy (5) for a specified α in advance. The statistic (3) can be expressed again as

$$T_i = \sqrt{\frac{n-1}{U_0 + U_i}} \times \frac{\bar{x}_i - (\bar{x}_0 + ia)}{\sqrt{\frac{\sigma^2}{n}}} = \sqrt{\frac{n-1}{U_0 + U_i}} \times V_i. \tag{6}$$

Here we can denote $Z_0 = \bar{x}_0/\sqrt{\sigma^2/n} \sim N(0,1)$ and express the probability density function as $h(z_0)$. It is difficult to decide a critical value in (5), then we consider the conditional distribution given $Z_0 = z_0^*, U_0 = u_0^*, U_i = u_i^*$ under $H_0^{(i)} (i = 1, \ldots, K)$. Then the statistic (6) is denoted by

$$T_i^* = \sqrt{\frac{n-1}{u_0^* + u_i^*}} \left(\frac{\bar{x}_i - ia}{\sqrt{\frac{\sigma^2}{n}}} - z_0^* \right) = \sqrt{\frac{n-1}{u_0^* + u_i^*}} \times V_i^*, \quad i = 1, \ldots, K, \tag{7}$$

where $V_i^* = (\bar{x}_i - ia)/\sqrt{\sigma^2/n} - z_0^* \sim N(-z_0^*, 1)$. Then

$$\begin{aligned}
1 - \alpha &= \Pr(|T_1| \le r, \ldots, |T_K| \le r) \\
&= \int_{-\infty}^{\infty} \int_0^{\infty} \cdots \int_0^{\infty} f(|T_1^*| \le r, \ldots, |T_K^*| \le r| \\
&\qquad\qquad U_0 = u_0^*, \ldots, U_K = u_K^*, Z_0 = z_0^*) \\
&\quad \times g(u_0^*) \times \cdots \times g(u_K^*) h(z_0^*) du_0^* \cdots du_K^* dz_0^* \\
&= \int_{-\infty}^{\infty} \int_0^{\infty} \left(\int_0^{\infty} \int_{-r\sqrt{\frac{u_0^*+u^*}{n-1}}}^{r\sqrt{\frac{u_0^*+u^*}{n-1}}} f(v^*) g(u^*) dv^* du^* \right)^K g(u_0^*) h(z_0^*) du_0^* dz_0^*,
\end{aligned} \tag{8}$$

where $f(v^*)$ is the probability density function of $N(-z_0^*, 1)$ and $g(u^*)$ is the probability density function of $\chi^2(n-1)$.

First, we give the values of K, n and α, then we can decide r by the multiple integration (8) so as to satisfy $1 - \alpha$.

4 Power of Test and Necessary Sample Size

We first define the power of the test π_1 under the configuration of μ_0, μ_1 as

$$\pi_1 = \Pr(|T_1| > r | \mu_1 \ne \mu_0 + a) = \int_{-\infty}^{-r} f(t_1) dt_1 + \int_r^{\infty} f(t_1) dt_1, \tag{9}$$

where $f(t_1)$ is the probability density function of T_1 and where T_1 is distributed according to noncentral t distribution with $2(n-1)$ degrees of freedom and non-centrality parameter $\{\mu_1 - (\mu_0 + a)\}/\sqrt{2\sigma^2/n}$. Similarly, we define the power of the test π_i under the configuration of $\mu_0, \ldots, \mu_i$ as

$$
\begin{aligned}
\pi_i &= \Pr(|T_1| \le r, \ldots, |T_{i-1}| \le r, |T_i| > r | i = \min\{i : \mu_i \ne \mu_0 + ia\}) \\
&= \int_{-\infty}^{\infty}\int_0^{\infty}\left(\int_0^{\infty}\int_{-r\sqrt{\frac{u_0^*+u^*}{n-1}}}^{r\sqrt{\frac{u_0^*+u^*}{n-1}}} f(v^*)g(u^*)dv^*du^*\right)^{i-1} \\
&\quad \times \int_0^{\infty}\left(\int_{-\infty}^{-r\sqrt{\frac{u_0^*+u^*}{n-1}}} f(v_i^*)g(u^*)dv_i^* + \int_{r\sqrt{\frac{u_0^*+u^*}{n-1}}}^{\infty} f(v_i^*)g(u^*)dv_i^*\right)du^* \\
&\quad \times g(u_0^*)h(z_0^*)du_0^*dz_0^*, \quad i = 2, \ldots, K,
\end{aligned}
\tag{10}
$$

where

$$
\begin{aligned}
Z_0 &= \frac{\bar{x}_0}{\sqrt{\frac{\sigma^2}{n}}} \sim N\left(\frac{\mu_0}{\sqrt{\frac{\sigma^2}{n}}}, 1\right), \\
T_i^* &= \sqrt{\frac{n-1}{u_0^* + u_i^*}}\left(\frac{\bar{x}_i - ia}{\sqrt{\frac{\sigma^2}{n}}} - z_0^*\right) = \sqrt{\frac{n-1}{u_0^* + u_i^*}} \times V_i^*, \\
V_i^* &= \frac{\bar{x}_i - ia}{\sqrt{\frac{\sigma^2}{n}}} - z_0^* \sim N\left(\frac{\mu_i - ia}{\sqrt{\frac{\sigma^2}{n}}} - z_0^*, 1\right).
\end{aligned}
\tag{11}
$$

We propose the procedure to determine the necessary sample size at each stage in our sequential procedure:

(1) Give the values of a, σ^2 and specify the values of $K, \alpha, \mu_0, \ldots, \mu_K$ and π.
(2) Give a tentative sample size n_0.
(3) Decide a tentative critical value r_0 so as to satisfy (8) for a given n_0.
(4) Calculate $\pi^{(0)}$ so as to satisfy (10) by using r_0 in (3).
(5) If $\pi^{(0)}$ in (iv) is equal to a specified π, we adopt the tentative sample size n_0 as a necessary sample size n; otherwise we change n_0 and again go to (3).

Here we estimate the values of a and σ^2 based on the empirical data under the same situation.

5 A Simulation Study

The purpose of the simulation study is to investigate the characteristics of our multiple comparison procedure in terms of the power of the test and the necessary sample size under various configurations of population means. We first give various configurations of population means for $a = 1, 2$ in Tables 1 and 2 under the specify $\alpha = 0.05, K = 4, \sigma^2 = 1$. We suppose that two null hypotheses $H_0^{(3)}$ in Tables 1 and 2 have linear tendency on the configurations of population means, i.e.,

Table 1 The power of the test and the necessary sample size ($a = 1$)

Configuration	$\pi_i(n = 10)$	$\pi_i(n = 20)$	$n(\pi_i = 0.90)$
(0, 1, 2, 1)	0.929	0.962	9
(0, 1, 2, 2)	0.371	0.733	31
(0, 1, 2, 4)	0.371	0.733	31
(0, 1, 0, *)	0.944	0.979	9
(0, 1, 1, *)	0.379	0.744	30
(0, 1, 3, *)	0.379	0.744	30

($\alpha = 0.05$, $K = 4$, $\sigma^2 = 1$)

Table 2 The power of the test and the necessary sample size ($a = 2$)

Configuration	$\pi_i(n = 10)$	$\pi_i(n = 20)$	$n(\pi_i = 0.90)$
(0, 2, 4, 4)	0.929	0.962	9
(0, 2, 4, 5)	0.371	0.733	31
(0, 2, 4, 7)	0.371	0.733	31
(0, 2, 2, *)	0.944	0.979	9
(0, 2, 3, *)	0.379	0.744	30
(0, 2, 5, *)	0.379	0.744	30

($\alpha = 0.05$, $K = 4$, $\sigma^2 = 1$)

$(0, 1, 2, 3)$ for $a = 1$ and $(0, 2, 4, 6)$ for $a = 2$. Then we show the power of the test π_i for $n = 10, 20$ and the necessary sample size n at each stage for the specify $\pi_i = 0.90$ in Tables 1 and 2.

Here the symbol $*$ in the configuration shows that we can take an arbitrary value on the population mean. From Tables 1 and 2, we can show the following results:

(1) π_i at $n = 20$ is more powerful than π_i at $n = 10$.
(2) When the population mean at the MED level separates remarkably from the linear tendency, π_i takes large value.
(3) If the MED level among the configurations of population means appears at an early stage, π_i takes large value and n somewhat takes small value.
(4) Though the configurations in Tables 1 and 2 are different, the powers take same values. Because the gapes of population means from linear tendency at the changing stage take same values.

6 A Case Study

In concrete technology, it is known that an admixture is used in order to make the mortar that has workability during construct. Civil engineer is interested in a suitable amount of admixture on the consistency of mortar in the workability. Then we carry out the flow test to find a amount in which the effect of admixture appears with increasing admixture. In the experiment, we first fix the amounts of portland

Table 3 Mortar flow at the zeroth stage (admixture 0)

No.	1	2	3	4	5	Mean
Values (cm)	14.35	13.85	14.45	14.40	13.85	14.18

Table 4 Mortar flow at the first stage (admixture $C \times 0.4\%$)

No.	1	2	3	4	5	Mean
Values (cm)	13.80	14.30	14.20	13.70	14.45	14.09

cement ($C = 5$ kg), sand (7.5 kg, less than 5 mm) and water (1.5 kg), we assume five amounts of admixture (dose levels), i.e., 0, $C \times 0.4\%$, $C \times 0.8\%$, $C \times 1.2\%$, $C \times 1.6\%$. In the case study, we deal with the one-sided test, since it is clear that the response of the admixture gets bigger with increasing admixture. We repeatedly measure five observations on the extent of mortar flow for each dose level in the flow test. From the past experiments, it can be assume that the response of the admixture shows a linear tendency up to the second stage with increasing admixture. Thus we suppose the linear tendency $y = \mu_0 + a \times i = 14 + 0.4 \times i$, where y(response) shows a sample mean of five observations on the extent of mortar flow at each dose level. If we specify $\pi_3 = 0.90, \alpha = 0.05, \sigma^2 = 0.9$, we obtain a critical value $r = 2.63$ and a necessary sample size $n = 5$. We must determine whether we carry our the flow test at the next stage or not according to the result of the sequential test.

First step
We first test the difference between $\mu_0 + 0.4$ and μ_1 by using $r = 2.63$ from Tables 3 and 4. Then we continue to the second step by $t_1 = -2.47 < r = 2.63$.

Second step
We test the difference between $\mu_0 + 0.4 \times 2$ and μ_2 from Tables 3 and 5. Then we continue to the third step by $t_2 = -0.33 < r = 2.63$.

Third step
We test the difference between $\mu_0 + 0.4 \times 3$ and μ_3 from Tables 3 and 6. Because we reject $H_0^{(3)}$ at the third step by $t_3 = 11.61 > r = 2.63$, we terminate the sequential test.

After all, we can find the difference from the linear tendency at the third stage. Figure 1 shows the change of a response with increasing amounts of the admixture (stage).

7 Conclusions

In this study, we realized a multiple comparison procedure to sequentially find a changing point of the response with increasing dose levels in the dose finding test. Our approach is to construct a stepwise test to identify the MED level for showing

Table 5 Mortar flow at the second stage (admixture $C \times 0.8\%$)

No.	1	2	3	4	5	Mean
Values (cm)	15.75	15.85	14.15	13.95	14.50	14.84

Table 6 Mortar flow at the third stage (admixture $C \times 1.2\%$)

No.	1	2	3	4	5	Mean
Values (cm)	19.90	18.65	19.65	18.40	19.70	19.26

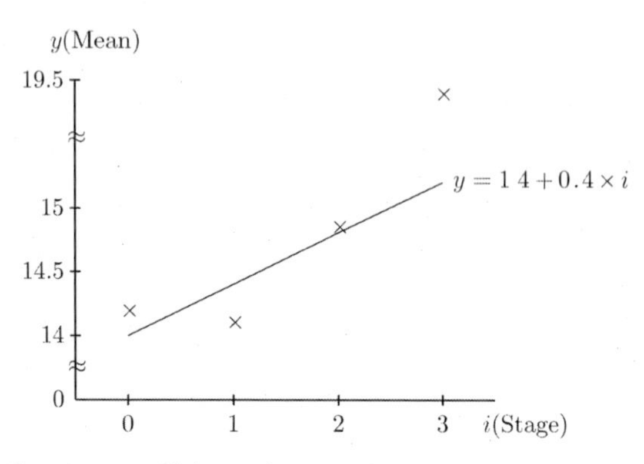

Fig. 1 A change of a response with increasing

the change from a linear tendency. First we proposed statistic and hypotheses at each step to incorporate the two-sided and one-sided step wise test. Since it is difficult to determine directly a critical value for satisfying a predefined type I familywise error rate, we presented a multiple integral involving the conditional distribution instead to the joint probability density function of several t statistics. By using the similar technique, we devised the multiple integral for the power of the test to first identify the MED level. However, it is unfeasible to calculate the numerical integral when the number of times on the integral increases.

From the simulation study, we could point out that the power of the test and the necessary sample size depend on the configurations of population means, especially due to the difference from the linear tendency at MED level and the position of the MED level in the configurations. In a case study, we could effectively find a suitable amount of admixture on the consistency of mortar in the workability by using our sequential procedure. However, there still remain the problem in which we must give the unknown $\mu_0, \ldots, \mu_K, a$ and the unknown variance σ^2 to calculate the power of the test and the necessary sample size.

References

BARTHOLOMEW, D. J. (1959a): A test of homogeneity for ordered alternatives. *Biometrika, 46*, 36–48.

BARTHOLOMEW, D. J. (1959b): A test of homogeneity for ordered alternatives. *Biometrika, 46*, 328–335.

CHEN, J. and GUPTA, A. K. (2000): *Parametric Statistical Change Point Analysis*. Birkhäuser, Boston.

DUNNETT, C. W. (1955): A multiple comparison procedure for comparing several treatments with a control. *Journal of the American Statistical Association, 50*, 1096–1121.

DUNNETT, C. W. and TAMHANE, A. C. (1991): Step-down multiple tests for comparing treatments with a control in unbalanced one-way layouts. *Statistics in Medicine, 10*, 939–947.

DUNNETT, C. W. and TAMHANE, A. C. (1992): A step-up multiple test procedure. *Journal of the American Statistical Association, 87*, 162–170.

DUNNETT, C. W. and TAMHANE, A. C. (1995): Step-up multiple testing of parameters with unequally correlated estimates. *Biometrics, 51*, 217–227.

MARCUS, R., PERITZ, E. and GABRIEL, K. R. (1976): On closed testing procedure with special reference to ordered analysis of variance. *Biometrika, 63*, 655–660.

SCHEFFÉ, H. (1953): A method for judging all contrasts in the analysis of variance. *Biometrika, 40*, 87–104.

TUKEY, J. W. (1953): *The Problem of Multiple comparisons*. Mimeographed monograph.

WILLIAMS, D. A. (1971): A test for differences between treatment means when several dose levels are compared with a zero dose control. *Biometrics, 27*, 103–117.

WILLIAMS, D. A. (1972): The comparison of several dose levels with a zero control. *Biometrics, 28*, 519–531.

Semi-supervised Clustering of Yeast Gene Expression Data

A. Schönhuth, I.G. Costa, and A. Schliep

Abstract To identify modules of interacting molecules often gene expression is analyzed with clustering methods. Constrained or semi-supervised clustering provides a framework to augment the primary, gene expression data with secondary data, to arrive at biological meaningful clusters. Here, we present an approach using constrained clustering and present favorable results on a biological dataset of gene expression time-courses in Yeast together with predicted transcription factor binding site information.

1 Introduction

Life on the biochemical level is driven by large molecules acting in concert following complex patterns in response to internal and external signals. Understanding these mechanisms has been the core question of molecular biology for the time since discovery of the DNA double helix. Ideally, one would like to identify detailed pathways of interaction. Unfortunately, this is often impossible due to data quality and the superposition of many such pathways in living cells. This dilemma led to the study of modules – sets of interacting molecules in one pathway – as identifying such modules is comparatively easy. In fact, clustering easily available mass data such as gene expression levels, which can be measured with DNA microarrays simultaneously for many genes is one approach for identifying at least parts of modules: for example co-regulated genes which show similar expression levels under several experimental conditions due to similarities in regulation.

The effectiveness of this approach is limited as we cluster based on observable quantities, the gene expression levels, disregarding whether the observed level can arise due to the same regulatory mechanism or not. Considering this information during the clustering should yield biologically more helpful clusters. Here we are

A. Schönhuth(✉)
ZAIK, Universität zu Köln, 50931 Cologne, Germany, E-mail: asa86@cs.sfu.ca

A. Okada et al. (eds.), *Cooperation in Classification and Data Analysis*,
Studies in Classification, Data Analysis, and Knowledge Organizations.
DOI: 10.1007/978-3-642-00668-5_16,

dealing with primary data, the gene expression levels, augmented with secondary data, for example transcription factor (TF) binding information.[1] Unfortunately, such secondary data is often scarce, in particular if we require high quality data.

Constrained clustering constitutes a natural framework. It is one of the methods exploring the gamut from unsupervised to supervised learning and it uses the secondary data to essentially provide labels for a subset of the primary data. Semi-supervised techniques have successfully been employed in image recognition and text classification (Lange et al. 2005; Lu and Leen 2005; Nigam et al. 2000). Hard constraints for mixture models (Schliep et al. 2004) were, to the best of our knowledge, the first application of constrained clustering in bioinformatics which showed the effectiveness of highest quality *must-link* or positive constraints indicating pairs of genes which should be grouped together. Here we use a soft version (Lange et al. 2005) which can cope with positive (*must-link*) and negative constraints (*must-not-link*) which are weighted with weights from [0, 1].

Constrained learning is used to estimate a mixture model where components are multi-variate Gaussians with diagonal covariance matrices representing gene expression time-courses. The secondary data consists of occurrences of transcription factor binding sites in upstream regions of yeast genes. Its computation is based on methods proposed in Rahmann et al. (2003) and Beer et al. (2004). The more transcription factor binding sites (TFBS) two yeast genes have in common, the more likely it is that they are regulated in a similar manner, which is reflected in a large positive constraint. Previously, we showed that even modest noise in the data used for building constraints actually will result in worse clustering solutions (Costa and Schliep 2006); the main contribution here is the careful construction of the secondary dataset and the method for evaluating the effectiveness of using constraints.

2 Methods

A mixture model (McLachlan and Peel 2000) is defined as

$$\mathbf{P}[x_i|\Theta] = \sum_{k=1}^{K} \alpha_k \mathbf{P}[x_i|\theta_k], \tag{1}$$

where $X = \{x_i\}_{i=1}^N$ is the set of (observed) data. The overall model parameters $\Theta = (\alpha_1, \dots, \alpha_K, \theta_1, \dots, \theta_K)$ are divided into the probabilities $\alpha_k, i = 1, \dots, K$ which add to unity for the model components $\mathbf{P}[x_i|\theta_k]$ and the $\theta_k, k = 1, \dots, K$, which describe the multi-variate Gaussians components of the mixture. One now aims at maximizing (1) by choosing an optimal parameter set Θ. This problem is routinely solved by the EM algorithm, which finds a local optimum for the above

[1] Transcription factors are essential for inhibiting or enhancing the production of proteins encoded in a gene.

function by involving a set of hidden labels $Y = \{y_i\}_{i=1}^N$, where $y_i \in \{1, \ldots, K\}$ is the component, which generates data point x_i. For details of the EM algorithm see Bilmes (1998).

In addition to the data x_i one is now given a set of positive respectively negative constraints w_{ij}^+ resp. $w_{ij}^- \in [0, 1]$, which reflect the degree of linking of a pair of data points $x_i, x_j, 1 \leq i < j \leq N$. The task is to integrate these constraints meaningfully and consistently into the EM routine. We will explain the essence of the solution proposed in Lange et al. (2005) and applied in Lu and Leen (2005) and Costa and Schliep (2006). Computation of the Q-function in each step of the EM-algorithm requires the computation of the posterior distribution $P[Y|X, \Theta]$ over the hidden labels y_i, where Θ is an actual guess for the parameters. By Bayes' rule we have

$$\mathbf{P}[Y|X, \Theta] = \frac{1}{Z} \cdot \mathbf{P}[X|Y, \Theta] \cdot \mathbf{P}[Y|\Theta], \tag{2}$$

where Z is a normalizing constant. The constraints are now incorporated by, loosely speaking, choosing as prior distribution $\mathbf{P}[Y|\Theta]$ the one, which is "most random" without that the constraints and that the prior probabilities α_k in Θ get violated. In other words, we choose the distribution, which obeys the *maximum entropy* principle and is called the *Gibbs* distribution (see Lange et al. 2005 for a theoretical setting and Lu and Leen 2005 for formulas and further details):

$$\mathbf{P}[Y|\Theta] = \frac{1}{Z} \prod_i \alpha_{y_i} \prod_{i,j} \exp[-\lambda^+ w_{ij}^+ (1 - \delta_{y_i y_j}) - \lambda^- w_{ij}^- \delta_{y_i y_j}], \tag{3}$$

where Z is a normalizing constant. The Lagrange parameters λ^+ and λ^- define the penalty weights of positive and negative constraints violations. This means that increasing λ^+, λ^- leads to an estimation, which is more restrictive with respect to the constraints. Note that computing (2) is usually infeasible and thus requires a *mean field approximation* (see again Lange et al. 2005 and Lu and Leen 2005 for details). Note, finally, that when there is no overlap in the annotations – more exactly, $w_{ij}^+ \in \{0, 1\}$, $w_{ij}^- \in \{0, 1\}$, $w_{ij}^+ w_{ij}^- = 0$, and $\lambda^+ = \lambda^- \sim \infty$ – we obtain hard constraints as the ones used in Schliep et al. (2005), or as implicitly performed in Pan (2006).

2.1 The Gene-TFBS-Matrix

The computational basis for the constraints is a binary valued incidence matrix, where the rows correspond to genes and the columns correspond to transcription factor binding sites (TFBS). A one indicates that, very likely, the TFBS in question occurs in the upstream region of the respective gene.

In a first step TFBS profiles were retrieved from the databases SCPD[2] and TRANSFAC.[3] In addition to consensus sequences for reported profiles we computed conserved elements in the upstream regions of the yeast's genes by means of the pattern hunter tool AlignACE.[4] In a second step we removed redundant patterns resulting in 666 putative TFBS sequence patterns. We then computed positional weight matrices (PWM) from these patterns by using G-C-rich background frequencies to contrast the patterns, following Rahmann et al. (2003).

With the PWMs we computed p-values for the occurrence of a TFBS in the upstream region of a gene by means of the following Monte Carlo approach. First, we generated 1,000 G-C-rich sequences of the length of the upstream sequences ($800bp$). We then computed a score for each of the 1,000 random sequences and each of the 666 PWMs by sliding a window of the length of the PWM in question over the sequence and adding up the values given by the PWM. We thus obtained, for each of the PWMs, a distribution of scores in sequences of length 800. We finally set a one in the Gene-TFBS-Matrix (GT-matrix) if the score of an upstream sequence of a gene (obtained by the same procedure as for the random sequences) was below a p-value of 0.001 compared to the distribution given through the random sequences. We note that we chose a very restrictive p-value as TFBS analysis usually is very easily corrupted by false positive hits (Rahmann et al. 2003; Claverie and Audic 1996) and false positives negate the benefits of constrained clustering.

2.2 Constraints

From the `GT-Matrix` we compute positive and negative constraints. We remind the reader that, by means of the `GT-Matrix` we have, for each of the genes, a binary valued vector of length 666. One is now tempted to, say, define the positive constraint between two genes to be proportional to the number of positions where the binary vectors of the two genes have a one in common (thus indicating that there is a transcription factor acting on both of the genes) and, likewise, to set the negative constraint to be proportional to the number of positions where exactly one of the genes has a one (thus indicating that there is a transcription factor which acts on one but not on both of the genes). Yet, although we expect seeing a one in only one of 1,000 genes in each of the columns of the matrix according to the p-value of 0.001, there are PWMs, which occur frequently (up to 90%) in the genes' upstream sequences. This indicates that there are heterogeneities in the upstream regions in general. It may also be due to the computation of the TFBSs as conserved elements of the upstream sequences themselves.

To address this we computed for each TFBS z the frequency of occurrence p_z within the genes and defined the positive (w_{ij}^+) and negative (w_{ij}^-) constraints for

[2] Saccharomyces cerevisiae promoter database, http://cgsigma.cshl.org/jian.

[3] The transcription factor database, http://www.gene-regulation.de.

[4] Motif finding algorithm, http://atlas.med.harvard.edu.

two genes i and j as follows. Let M_{iz} denote the `GT-Matrix` entry for gene i and TFBS z and set

$$w_{ij}^{+} := \gamma^{+} \cdot \#\{x \ : \ p_z^2 \leq 0.01, \ M_{iz} = M_{jz} = 1\}.$$

That is, w_{ij}^{+} is up to a scaling factor γ^{+}, the number of TFBSs, which occur with a p-value of 0.01 or less in both genes i and j. Similarly, we define

$$\begin{aligned} w_{ij}^{-} := \gamma^{-} \cdot (\#\{z \ : \ p_z(1-p_z) \leq 0.01, \ M_{iz} = 1, M_{jz} = 0\} \\ + \#\{z \ : \ p_z(1-p_z) \leq 0.01, \ M_{iz} = 0, M_{jz} = 1\}). \end{aligned}$$

2.3 Relevant Constraints

Constrained clustering profits from information of two datasets – the original, primary dataset and the secondary one, from which constraints are computed. When the influence of the secondary dataset is increased, cluster results change. To identify which constraints cause changes we computed the pairs of genes in one cluster, which were in the same cluster in the unconstrained clustering and in distinct clusters in the constrained case or vice versa. Lists of positive and negative constraints for pairs of genes identified ranked by constraint weight serve as the basis for further analysis. This way we identified the TFBSs which had the largest contribution to changes in the clustering.

3 Results

As in Costa and Schliep (2006) we used 384 yeast cell cycle gene expression profiles (YC5) for analysis. YC5 is one of the rare examples of a dataset where high quality labels are available for each gene as each of them is assigned to one of the five mitotic cell cycle phases. Because of the synchronicity of the profiles within one group (corresponding to one of the five phases), we opted for multivariate Gaussians with diagonal covariance matrices as components in the mixture model. We initialized the mixture estimation procedures by means of an initial model collection algorithm presented in Schliep et al. (2005). The clustering solution was obtained from the mixture by assigning each data point to the component of highest posterior probability.

3.1 Clustering Statistics

We estimated mixtures for varying values of the Lagrangian parameters λ^{+}, λ^{-}. Let TP resp. TN denote the amounts of pairs of genes correctly assigned to one resp.

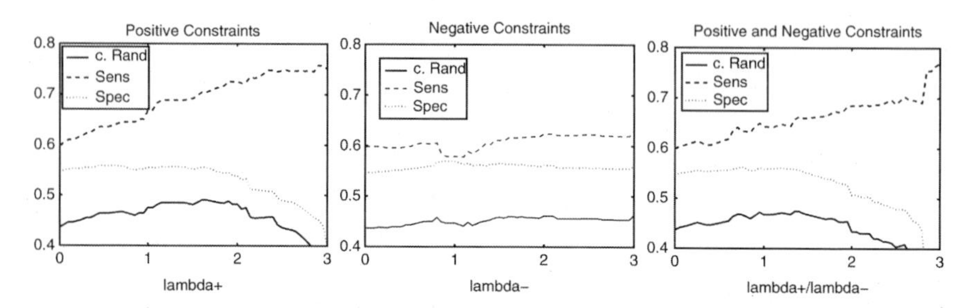

Fig. 1 We depict the CR, Spec and Sens with only positive (*left*), only negative (*middle*) and both positive and negative (*right*) constraints for increasing values of the Lagrangian parameters λ^+, λ^-

two clusters out of P resp. N many according to the true labels. Then, we computed $\text{Sens} = \frac{TP}{P}$ and $\text{Spec} = \frac{TN}{N}$, and the corrected Rand, which can be perceived as a significance level for the clustering of being distinct from a random distribution of the genes over the clusters, to monitor the effects of an increasing influence of the constraints (Fig. 1).

While the positive constraints improve sensitivity, the negative constraints slightly improve specificity. One also sees a considerable improvement of the corrected Rand for the addition of positive constraints and a slight improvement for the negative constraints. Taking into account both positive and negative constraints one sees improvements in all of the three statistics. However, there does not seem to be a synergy of the positive effects of the two kinds of constraints. This may be an indication for contradictions within the constraints and suggests some "contradiction purging" as a future area of research.

3.2 Gene Ontology Statistics

To validate the clustering quality from a biological point of view we compare the p-values from enrichment of Gene Ontology (GO) terms in a procedure similar to the one performed in Ernst et al. (2005). More specifically, we computed GO term enrichment using GOStat (Beissbarth and Speed 2004) for an unconstrained and a constrained ($\lambda^+ = \lambda^- = 1.35$) mixture estimation as described above. We selected all GO terms with a p-value lower then 0.05 in both clusterings and plotted the $-\log(p\text{-values})$ of these terms in Fig. 2.

We found smaller p-values for the constrained clustering and compile a list of GO Terms, which display high log-ratios in Table 1. The constrained case had 16 of such GO terms, 10 out of these are directly related to biological functions or cell compartments related to cell cycle (big dots[5] in Fig. 2 and GO terms in italic in Table 1). On the other hand, only five GO terms had a higher enrichment in the unconstrained case, all with a significant lower log ratio then in the constrained case.

[5] Due to overlap, only eight big dots are visible.

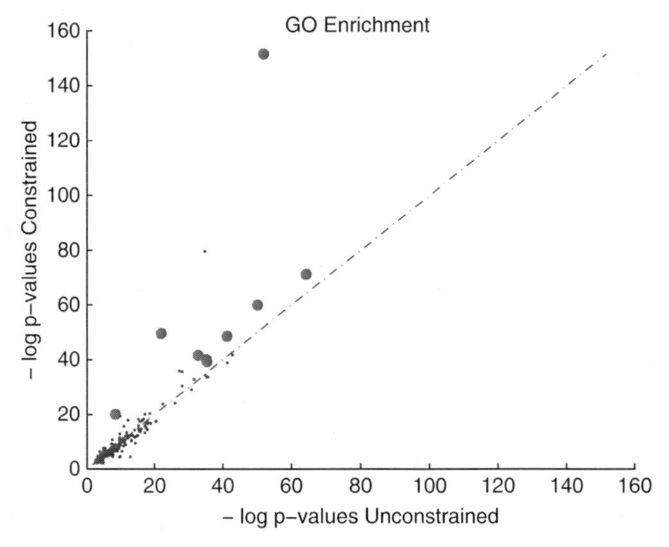

Fig. 2 Scatter plot comparing the GO Term enrichment of the unconstrained (x-axis) and constrained (y-axis) results. Points above the *diagonal line* indicate higher enrichment in the constrained case, while values below indicate higher enrichment in the unconstrained case

From those, the first four are related to chromatin structure and nucleosome, which is related to the S phase of cell cycle.

As described in Sect. 2.3 we computed the constraints which had a relevant impact on the clustering statistics. We found (not shown) that the TFBS data particularly helped correctly classifying genes, which belong to cell cycle phases *late G1* and *S* which is consistent with the gene expression time-course dataset used. Further manual analysis of the relevant constraints and investigation of the TFBSs involved will likely provide insights in mechanisms which are not discoverable from gene expression alone.

4 Conclusion

Constrained clustering is a very useful tool for analyzing heterogeneous data in molecular biology, as there is often an abundant primary data source available (e.g., gene expression, sequence data) which can be made much more useful by integration of high-quality secondary data. However, as the results by Costa and Schliep (2006) show, constrained clustering cannot be applied straight-forwardly even to secondary data sources which are routinely used for biological validation of clustering solutions. Point in case: the *predicted* TFBS information used here improves results whereas the *experimental* chip-on-chip data used by Costa and Schliep (2006) does not. This is likely due to higher error rates in the experimental data and a lack of quality measure for each individual experiment, which precludes filtering

Table 1 List of GO Terms for which the log ratio of the p-values is higher then 4.0 (or $|\log((p\text{-values const.})/(p\text{-values unconst.}))| > 4.0$). Positive ratios indicate a higher relevance of the term in a cluster from the constrained case, while negative ratios indicates higher relevance in a cluster from the unconstrained case

GO Term ID	GO Term	p-value log ratio
GO:0005694	*Chromosome*	99.6581
GO:0009719	Response to endogenous stimulus	44.9090
GO:0000278	*Mitotic cell cycle*	27.6137
GO:0003677	*DNA binding*	11.7053
GO:0044427	*Chromosomal part*	9.7880
GO:0007010	Cytoskeleton organization and biogenesis	9.7352
GO:0000228	*Nuclear chromosome*	8.9036
GO:0043232	Intracellular non-membrane-bound organelle	8.6498
GO:0043228	Non-membrane-bound organelle	8.6498
GO:0044454	*Nuclear chromosome part*	7.5673
GO:0007049	*Cell cycle*	7.4107
GO:0006259	*DNA metabolism*	6.9792
GO:0044450	Microtubule organizing center part	5.6984
GO:0006281	*DNA repair*	4.9234
GO:0007017	Microtubule-based process	4.7946
GO:0006974	*Response to DNA damage stimulus*	4.0385
GO:0000786	Nucleosome	−8.3653
GO:0000788	Nuclear nucleosome	−8.3653
GO:0000790	Nuclear chromatin	−5.1417
GO:0000785	Chromatin	−5.1333
GO:0016043	Cell organization and biogenesis	−4.8856

on quality. Noise reduction in constraints, resolution of conflicts between positive and negatives constraints and measure of constraint relevance are open questions which need to be addressed.

References

BEER, M. A. and TAVAZOIE, S. (2004). Predicting gene expression from sequence. *Molecular Biology of the Cell, 117,* 185–198.

BEISSBARTH, T. and SPEED, T. P. (2004): GOStat: find statistically overrepresented Gene Ontologies within a group of genes. *Bioinformatics, 20:*1464–1465.

BILMES, J. A. (1998): A gentle tutorial of the EM algorithm and its application to parameter estimation for Gaussian mixture and hidden Markov models. *Technical Report TR-97-021, International Computer Science Institute, Berkeley, CA.*

CLAVERIE, J. M. and AUDIC, S. (1996): The statistical significance of nucleotide position-weight matrix matches. *CABIOS, 12(5):*431–439.

COSTA, I. and SCHLIEP, A. (2006): On the feasibility of heterogeneous analysis of large scale biological data. In *ECML/PKDD Workshop on Data and Text Mining for Integrative Biology, pages* 55–60.

ERNST, J., NAU, G. J. and BAR-JOSEPH, Z. (2005): Clustering short time series gene expression data. *Bioinformatics, 21:* i159–i168.

LANGE, T., LAW, M. H. C., JAIN, A. K. and BUHMANN, J. M. (2005): Learning with constrained and unlabelled data. In *IEEE Conference on Computer Vision and Pattern Recognition (CVPR'05), volume 1, pages* 731–738.

LU, Z. and LEEN, T (2005): Semi-supervised learning with penalized probabilistic clustering. In Lawrence K. Saul, Yair Weiss, and Léon Bottou, editors, *Advances in Neural Information Processing Systems 17, pages 849–856.* MIT, Massachusetts.

MCLACHLAN, G. and PEEL, D. (2000): *Finite Mixture Models.* Wiley Series in Probability and Statistics. Wiley, New York.

NIGAM, K., McCALLUM, A. K., THRUN, S. and MITCHELL, T. (2000): Text classification from labeled and unlabeled documents using EM. *Machine Learning, 39(2/3):*103–134.

PAN, W. (2006): Incorporating gene functions as priors in model-based clustering of microarray gene expression data. *Bioinformatics, 22(7):*795–801.

RAHMANN, S., MUELLER, T. and VINGRON, M. (2003): On the power of profiles for transcription factor binding site detection. *Statistical Applications in Genetics and Molecular Biology, 2(1):*7.

SCHLIEP, A., STEINHOFF, C. and SCHÖNHUTH, A. (2004): Robust inference of groups of genes using mixtures of HMMs. *Bioinformatics, 20(suppl 1):*i283-i289.

SCHLIEP, A., COSTA, I. G., STEINHOFF, C. and SCHÖNHUTH, A. (2005): Analyzing gene expression time-courses. *IEEE/ACM Transactions on Computational Biology and Bioinformatics, 2(3):*179–193.

Event Detection in Environmental Scanning: News from a Hospitality Industry Newsletter

R. Wagner, J. Ontrup, and S. Scholz

Abstract Recognizing current and future developments in the business environment is a major challenge in business planning. Although modern information retrieval (IR) technologies become more and more available in terms of commercial software, sophisticated technologies are rarely applied in environmental scanning. Moreover, modern text classification methodologies are being applied separately rather than being linked with other IR approaches to gain a substantial surplus in managerial information assessment. In this paper we combine the Hierarchically Growing Hyperbolic SOM with the Information Foraging Theory for an appraisal and selection of relevant documents. We demonstrate this approach in detecting important up-and-coming developments in the business environment using a stream of documents from a newsletter service for the hospitality industry.

1 Introduction

The ability to scan and to identify key information efficiently and effectively is one of the most crucial skills for business managers who want to succeed in today's information-intensive environments (Huang 2003). Herbert A. Simon (quoted by Varian 1995) highlights this problem in the following way: "What information consumes is rather obvious: it consumes the attention of its recipients. Hence a wealth of information creates a poverty of attention, and a need to allocate that attention efficiently among the over abundance of information sources that might consume it". Summarizing these shortcomings we claim that one of the exigent challenges in business information technology is to integrate the concepts of clustering, classification and pattern recognition. In this paper we:

- Outline the structuring of documents and event detection by means of clustering using the Hierarchically Growing Hyperbolic SOMs

R. Wagner(✉)
Dialog Marketing Competence Center, University of Kassel, 34125 Kassel, Germany,
E-mail: rwagner@wirtschaft.uni-kassel.de

A. Okada et al. (eds.), *Cooperation in Classification and Data Analysis*,

Studies in Classification, Data Analysis, and Knowledge Organizations.
DOI: 10.1007/978-3-642-00668-5_17,

- Demonstrate the evaluation and assessment of documents in the structured text corpus by dint of an Information Foraging Theory based approach

The term environmental scanning refers to the way in which managers study their relevant business environment. In the hospitality industry as well as in almost all other industries the environmental scanning activities are hampered by manager's information overload due to the increase of relevant documents and messages emitted by an steadily growing number of information sources (Scholz and Wagner 2005). The need for spending efforts, time, and resources in these activities arises from the competition on recognizing new developments before a competitor does. Superior business strategies and tactics need to be adjusted in advance to the new developments.

Since information quality constitutes a major problem, the sheer plethora of uncensored sources which embody information on new developments in the business environment makes the WWW an ideal environment for detecting changes crucial for future success. Important features of this virtual information structure are:

(a) The temporal information (time stamp) contained in documents, such as news feeds and tickers (e.g., Reuters or Deutsche Presse Agentur), chatroom messages, web forums and blogs as well as system log messages.
(b) The neighborhood of individual messages including the absence or presence of the originator information.

The remainder of this paper is structured as follows: In Sect. 2 we provide a brief data description. Since data of this type is typically available to managers in various industries, the scope of our paper is not limited to the application example. Subsequently, in Sect. 3 we outline our methodologies and apply them to the documents at hand. Finally, we discuss the results of combining these methodologies.

2 Data Description and Preprocessing

To demonstrate our methodology we draw on the example of 2,314 documents obtained from an internet-based hospitality industry newsletter. The documents were posted in a period from July 2002 to May 2003. Three major threats had impact on the hospitality industry: The Bali attacks in October 2002, the start of the Iraq war in March 2003, and the outbreak of the SARS epidemic in April 2003. Postings to this newsletter were not restricted to messages and appraisals concerned with these major events. They also cover mergers and acquisitions within the industry, releases of official communiques with relevance to the industry as well as company reports and conference announcements, etc. Approximately 12 messages were posted per day. Figure 1 exemplifies the nature of these messages.

Using Porter's stemming algorithm a classical bag-of-words representation with 7,457 entries is computed. In addition to this, we rely on the term frequency-inverted

Best Western Announces Fourth Quarter Results
Further solidifying its position as THE WORLD'S LARGEST HOTEL CHAIN Best Western's worldwide development team finished the year strong by signing 83 new hotels to the brand during the fourth fiscal quarter ending Nov. 30. During the quarter, 42 hotels were added in North America and 41 were signed internationally. [...] In October, the company also opened the Best Western Greenwood Inn and Suites in Beaverton, Ore.[...]

Fig. 1 Example of newsletter messages

document frequency weighting scheme and cosine similarity measures to compute distances between the documents.

3 Methodology

3.1 Hierarchically Growing Hyperbolic Maps

In order to generate a structured information source from an unstructured text collection we suggest the application of the Hierarchically Growing Hyperbolic Map (H^2SOM) (Ontrup and Ritter 2005) which is an extension to the Self-Organizing Map as discussed by Kohonen (2001). Whereas standard SOM approaches most prominently use a regular grid in 2D Euclidean space as their map canvas, we embed the nodes of the H^2SOM in hyperbolic space which is characterized by being negatively curved. The peculiar geometric properties make the hyperbolic plane $I\!H^2$ an ideal candidate for the embedding of large hierarchies (Lamping and Rao 1994). For its display on a flat 2D screen one may choose the projection on the Poincaré Disk that has a number of beneficial properties for visualization. First, it is locally shape preserving with a strong fish-eye effect: the origin of $I\!H^2$ – corresponding to the fovea – is mapped almost faithfully, while distant regions become exponentially "squeezed". Second, the model allows an elegant translation of the fovea by means of Möbius transformations, enabling the user to selectively focus on interesting parts of the map while still keeping a coarser view of its surrounding context. For more details on the predecessor of the H^2SOM, the HSOM and its applications see, e.g., Ritter (1999) or Walter et al. (2003).

3.1.1 Network Architecture

(1) The network starts with a ring of n_b equilateral hyperbolic triangles centered at the origin of $I\!H^2$ as shown in Fig. 2a. The $n_b + 1$ vertices of the triangles form the network nodes of the first level of the hierarchy. (2) During a growth step we can expand each node in the periphery of the existing network by surrounding it with

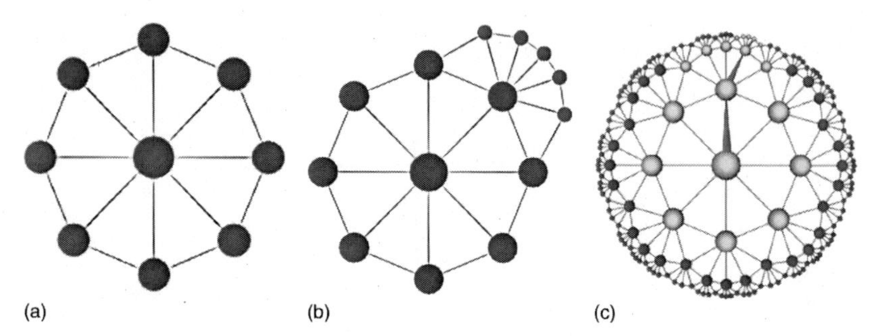

Fig. 2 (**a**) The nodes at the vertices of – in this case $n_b = 8$ – equilateral triangles form the first level in the hierarchy of the H^{2}SOM. (**b**) When a node meets a growth criterion, we generate a set of $n_b - 3$ children nodes. In (**c**) the search path and the set of nodes visited during a best match search are highlighted within a fully expanded hierarchy of three levels. Note, that the nodes are placed on an equidistant grid in hyperbolic space, only the projection on the 2D Poincaré Disk introduces a strong distortion of distances at the perimeter of the disk

the vertices of additional $n_b - 2$ equilateral triangles. By repeating (2) for all border nodes, we can generate a new "ring" level of the hierarchy. The "branching" factor $n_b > 6$ determines how many nodes are generated at each level and how "fast" the network is reaching out into the hyperbolic space.

3.1.2 Learning and Growing Procedure

The training of the hierarchical network largely follows the traditional SOM approach. To each node a we attach a reference vector $\mathbf{w}_a$ projecting into the input data space X. The initial configuration with the first n_b nodes is trained in the "usual" SOM way (Kohonen 2001) with one distinct difference: The node distances are computed using their 2D positions $\mathbf{z}_a \in \mathbb{C}$ in the complex Poincaré Disk $|\mathbf{z}| \leq 1$: $d_{a,a^*} = 2(|\mathbf{z}_a - \mathbf{z}_{a^*}|/|1 - \mathbf{z}_a\bar{\mathbf{z}}_{a^*}|)$ (Coxeter 1957).

After the training of a ring with fixed training intervals we repeatedly evaluate for each node an expansion criterion. In our experiments we have so far used the node's quantization error. If a given threshold Θ_{QE} for a node is exceeded, that node is expanded as described in step (2) above. After the expansion step where all nodes meeting the growth criterion were expanded, all reference vectors from the previous hierarchies become fixed and adaptation "moves" to the nodes of the new structural level. In contrast to the standard SOM, the H^{2}SOM has not to carry out a global search for determining the winner node a^*, but does instead perform a fast tree search, taking as the search root the initial center node of the growth process and following the natural hierarchical structure in the hyperbolic grid as indicated in Fig. 2c. This approximative search allows to train self-organized maps with a quality on par or even better that standard Euclidean SOMs, but several magnitudes faster (Ontrup and Ritter 2005).

3.1.3 Application

The H^2SOM is utilized in the following manner: The top level hierarchy of the network is trained with the complete set of documents. This initial map then partitions the collection into a set of subclusters. Nodes are expanded and the next structural level is trained by performing a best-match search along the already build up hierarchy. In this way, the growing network splits the document collection into more and more subclusters of decreasing size which are arranged in a topology preserving manner on the hierarchical grid of the hyperbolic plane. After the nodes have reached their prescribed quantization error or a previously defined maximal depth in the hierarchy all nodes of the network get labelled. As labels are chosen those terms from the bag-of-words model which correspond to the maximal values in the prototype vectors attached to the nodes. By dragging the mouse, the fovea of the map can be moved continuously in hyperbolic space, such that a desired level of detail can be selected in a "focus and context"-like manner. Depending on the position of the fovea, the keywords are dynamically adjusted such that the currently selected area of interest is labelled with high density, whereas the surrounding context gets labelled with fewer keywords.

3.1.4 Event Detection in Document Streams

By attaching to each neuron of the H^2SOM a time dependent activation potential defined as $a_i(t) = \beta\, a_i(t-1) + \mathcal{S}_i(t)$, with

$$\mathcal{S}_i(t) = \begin{cases} 1, & \text{if } i \text{ is best-match node at arrival time } t, \\ 0, & \text{otherwise,} \end{cases}$$

where β is a decay factor controlling the amount of leakage, each node of the H^2SOM acts like a *leaky integrator*. As news items "flow" in, the neuron activities of the corresponding best match nodes in the hierarchy increase. At times with no news coverage, node activations decrease again. By continuously mapping the incoming data stream to the H^2SOM, a "movie" of news activities can then be generated.

In Fig. 3 two screenshots show how the hierarchical organization of the documents together with an animation of time dependent news activities can be used to gain insights on "where" and "when" in the document collection potentially interesting events occur.

3.2 *Information Foraging Theory*

While H^2SOM clustering leads to a meaningful structure of the information space the resulting (sub)clusters often still include an amount of news messages that cannot be read entirely–even when the cluster is regarded as relevant. In order to reduce

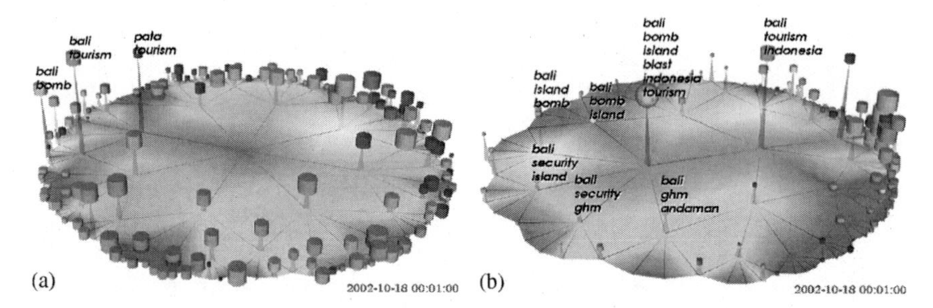

Fig. 3 In mid October 2002 a significant peak in the news activity can be observed in a subcluster labelled "bali, tourism". In (**b**) the user has moved the focus to a deeper structural level of that subcluster where the keywords suggest that a bomb blast on the island of Bali has generated this news peak

the amount of information to a manageable size we apply the Information Foraging Theory (IFT).

IFT refers to activities associated with assessing and extracting information in a given information space (Pirolli and Card 1999). The basic premise of IFT is that information seeking tasks are akin to animal behavior when searching for food. The concepts of *information diet* and *information gain* are indispensable in the theoretic understanding of IFT:

The forager has to optimize his limited time resources for an efficient allocation between information search and information consumption. Detailed descriptions of the basic principles of *information diet* composition in the domain of environmental scanning are given in Scholz and Wagner (2006), as well as (Decker et al. 2005). In order to select the relevant messages from a continuous stream the *information gain* of a message is appraised by means of its relevance in a given context. The relevance of information is depending on specific information needs which are directly related to the manager's mental model.

Thereby the relevance of a message is not evaluated by reading each message entirely, but by pre-estimating the value of a document by means of proximal cues. For a computable representation of the messages we apply the aforementioned classical bag-of-words model. The assessment of the information scent by means of language expressions which embody proximal cues is modeled by a spreading activation network.

The activation of a memory A_k which is fed with a proximal cue (a language expression k) can be computed by means of a Bayesian prediction of the relevance:

$$A_k = C_k + \sum_l W_l S_{kl}. \tag{1}$$

The base-level activation C_k (and W_l) of different language terms k and l varies due to the previous knowledge and information needs of the manager. We denote the set of relevant terms that constitute the manager's information needs as the information structure $\mathcal{Q}$. Language expressions which refer to relevant concepts in the managers' appreciation of information needs ought to have high association strengths S_{kl}, whereas low strengths denote dissociated terms. The association strength of a proximal cue (language term or concept of the mental model) is computed with the help of the following three equations (Pirolli and Card 1999):

$$C_k = \ln\left(\frac{p(k)}{p(\bar{k})}\psi\right) \quad \forall\ k = 1, \ldots, K, \tag{2}$$

$$W_l = \ln\left(\frac{p(l)}{p(\bar{l})}\psi\right) \quad \forall\ l = 1, \ldots, k-1, k+1, \ldots, K, \tag{3}$$

$$S_{kl} = \ln\left(\frac{p(l|k)}{p(l|\bar{k})}\psi\right) \quad \forall\ l \neq k, \tag{4}$$

with $p(k)$ and $p(l)$ denoting the probabilities of the terms k (and l respectively) occurring in the information space $\mathcal{I}$. The probabilities $p(\bar{k})$ and $p(\bar{l})$ reflect the "absence" of these terms in the information environment. The posterior probabilities $p(l|k)$ and $p(l|\bar{k})$ denote the conditional probability of the occurrence of term l in the context of term k and the conditional probability of term l occurring in a context that does not contain word term k. Positive values are ensured by the normalizing constant ψ.

The resulting information scent arising from the association strengths of language expressions that are given in the manager's information structure $\mathcal{Q}$ and occur in message m_{ij} is used to assess the information gain of each message:

$$g_{ij} = \exp\left(\frac{\sum_{k \in \mathcal{Q} \cap m_{ij}} A_k}{Z}\right) \quad \forall m_{ij} \in \mathcal{I}, \tag{5}$$

where Z is a scaling factor which is estimated on an a priori characterization of the information environment. The index i refers to the information patch (e.g., newsgroup or cluster) from which a message j has been obtained. In our application the patches i are made up by all messages assigned to the same best matching node i due to the H^2SOM clustering.

When applying IFT to the cluster comprising messages about the bombing attacks in Bali (cf. Fig. 3) the information diet comprises of four messages from 97 only. The selected messages represent the news bursts about the Bali attack in an appropriate manner: two messages which are included in the information diet are the first news published within the first 24 h after the terror attack took place. Another two messages represent a second news burst delivered in January which deals with the economic impacts of the terror attacks on the Asian hospitality industry. Thus, both the event in itself and also the ensuing discussion about its impact on the hospitality industry can be traced when reading only these four messages.

4 Discussion and Conclusions

Combining both methods presented herein offers several advantages for managers scanning and monitoring their business environment. In a first step, the H^2SOM may be utilized to generate a hierarchically structured information source suitable for an interactive visual exploration. The annotation of the information patches and the spatial relation to neighboring patches on the "document landscape" provide an intuitive understanding of the underlying data structure. Once a possibly interesting topic is identified by a manager, the documents relevant to this topic are mapped to a node in the H^2SOM. This set may then be regarded as an information structure $\mathcal{Q}$ defining the manager's information needs in the sense of the IFT. The example of the terror attacks on Bali indicates, that by "digesting" the selected (sub)cluster, the IFT is able to significantly reduce the amount of information the manager actually has to read without loosing important information with respect to $\mathcal{Q}$. Similar results on the text corpus (not reported here due to space limitations) were obtained with information on hotel mergers, the outbreak of the SARS virus or the war in Iraq. A further extension of our methodological outline could be a monitoring function based on statistical testing of changes in word associations in the course of time.

References

COXETER, H.S.M. (1957): *Non Euclidean Geometry*, University of Toronto Press, Toronto.

DECKER, R., WAGNER, R., and SCHOLZ, S.W. (2005): Environmental Scanning in Marketing Planning. *Marketing Intelligence and Planning, 23*, 189–199.

HUANG, A.H. (2003): Effects of Multimedia on Document Browsing and Navigation: An Exploratory Empirical Investigation. *Information and Management, 41*, 189–198.

KOHONEN, T. (2001): *Self-Organizing Maps*, 3rd edn. Springer, Berlin.

LAMPING, J. and RAO, R. (1994): Laying Out and Visualizing Large Trees Using a Hyperbolic Space. In: *Symposium on User Interface Software and Technology.* ACM, New York, 13–14.

ONTRUP, J. and RITTER, H. (2005): A Hierarchically Growing Hyperbolic Self-Organizing Map for Rapid Structuring of Large Data Sets. In: *Proceedings of the 5th Workshop on Self-Organizing Maps (WSOM 05).*

PIROLLI, P. and CARD, S. (1999): Information Foraging. *Psychological Review, 106*, 643–675.

RITTER, H. (1999): Self-Organizing Maps in Non-Euclidian Spaces. In: E. Oja and S. Kaski (Eds.): *Kohonen Maps*. Elsevier, Amsterdam, 97–110.

SCHOLZ, S.W. and WAGNER, R. (2005): The Quality of Prior Information Structure in Business Planning? – An Experiment in Environmental Scanning. In: H. Fleuren, D. den Hertog, and P. Kort (Eds.): *Operations Research Proceedings 2004*. Springer, Berlin, 238–245.

SCHOLZ, S.W. and WAGNER, R. (2006): Autonomous Environmental Scanning in the World Wide Web. In: B.A. Walters and Z. Tang (Eds.): *IT-Enabled Strategic Management: Increasing Returns for the Organization.* Idea, Hershey, 213–242.

VARIAN, H.A. (1995): The Information Economy – How Much Will Two Bits be Worth in the Digital Marketplace? *Scientific American, Sept.*, 200–201.

WALTER, J., ONTRUP, J., WESSLING, D., and RITTER, H. (2003): Interactive Visualization and Navigation in Large Data Collections using the Hyperbolic Space. In: *Proceedings of the Third IEEE International Conference on Data Mining.*

Part III
Applications in Clustering and Visualization

External Asymmetric Multidimensional Scaling Study of Regional Closeness in Marriage Among Japanese Prefectures

A. Okada

Abstract The purpose of the present study is to disclose the relationships among Japanese prefectures in the marriage by using asymmetric multidimensional scaling. The analysis was done by external multidimensional scaling where the location of each prefecture was given externally. The data consist of frequencies of the marriage between any two prefectures from 2001 to 2005. The two-dimensional solution was derived which represented asymmetric relationships among prefectures. The result shows that the female of prefectures neighboring Tokyo has the smaller tendency of getting married to the male from the other prefectures.

1 Introduction

Relationships among Japanese prefectures have been studied by several researchers (e.g., Slater 1976). These studies focus their attention on the relationships among Japanese prefectures in the university enrollment flow; based on the number of high school graduates from a prefecture who enter into universities in another prefecture. The relationships among the prefectures are inevitably asymmetric. Some researchers paid their attention on the asymmetry of relationships among Japanese prefectures in the university enrollment flow (e.g., Okada and Imaizumi 2005; Okada and Iwamoto 1995, 1996). In the present study relationships among Japanese prefectures in the marriage are dealt with. The relationships among Japanese prefectures in the marriage seem also asymmetric, because it has been said that the male living in prefectures in rural farming areas are somewhat more difficult to find the match than those living in prefectures in the urbanized area. The data, the frequency of the marriage of the female from a prefecture and the male from another

A. Okada
Graduate School of Management and Information Sciences, Tama University, 4-1-1 Hijirigaoka, Tama-shi, Tokyo 206-0022, Japan, E-mail: okada@tama.ac.jp

A. Okada et al. (eds.), *Cooperation in Classification and Data Analysis*,
Studies in Classification, Data Analysis, and Knowledge Organizations.
DOI: 10.1007/978-3-642-00668-5_18,

prefectures for 47 Japanese prefectures, were analyzed by the external asymmetric multidimensional scaling.

2 The Data

The data analyzed in the present study consist of frequencies of the marriage between any combination of two prefectures; one prefecture from where the female comes and the other from where the male comes. The data form a 47×47 table where the (j, k) element of the table shows the frequency of the marriage which the female comes from prefecture j and the male comes from prefecture k. Because the frequency of the marriage which the female comes from prefecture j and the male comes from prefecture k is not necessarily equal to the frequency of the marriage which the female comes from prefecture k and the male comes from prefecture j, the 47×47 table is not necessarily symmetric, and the data are one-mode two-way asymmetric proximities.

The data were collected by a marriage matchmaking service company for the 5 years from 2000 to 2005 (Tanaka 2005). The total frequency of the marriage in the 5 years is 11,115. The (j, k) element of the table is the frequency of the marriage that (a) the female from prefecture j and (b) the male from prefecture k married, and they (c) quitted the service of the company. We regard the (j, k) element of the table as the similarity from prefectures j to k.

3 The Method

The 47×47 table of the frequency of the marriage among 47 Japanese prefectures was analyzed by one-mode two-way asymmetric multidimensional scaling (Okada and Imaizumi 1987) externally. Like the usual one-mode two-way asymmetric multidimensional scaling or internal one-mode two-way asymmetric multidimensional scaling, each prefecture (object) is represented as a point and a circle in the two-dimensional space (a sphere in the three-dimensional space, a hypersphere in more than three-dimensional space) centered at that point representing the prefecture in a multidimensional Euclidean space.

The external analysis was done by using the geographical location or the map of 47 prefectures as the externally given two-dimensional configuration of objects (prefectures). The geographic location of each prefecture, or the two-dimensional configuration of 47 prefectures, is given by the longitude (the horizontal dimension) and the latitude (the vertical dimension) of the capital of the prefecture (Okada and Imaizumi 2005). The reason for executing the external analysis is that the geographical location of the prefectures must have a great deal of effect on the relationships among prefectures. The external analysis of one-mode two-way asymmetric multidimensional scaling derives the radius of each prefecture. The radius of a prefecture represents the (relative) asymmetry of the prefecture. The data were also analyzed by the asymmetric cluster analysis (Okada 2000; Okada and Iwamoto 1996) to examine the effect of the geographical location of prefectures.

4 The Analysis

The analysis of the 47 × 47 table of the frequency of the marriage among 47 Japanese prefectures was done in the two-dimensional space. Because the external analysis was done by giving the two-dimensional configuration of 47 Japanese prefectures, the analysis was done only in the two-dimensional space. Analyses in the spaces of several kinds of different dimensionality, usually done in the internal analysis (the usual analysis utilizing asymmetric multidimensional scaling without using any external information or the configuration of prefectures in the present application) , are not needed in the external analysis. By the external analysis, only the radius for each of the 47 prefectures was derived for the given configuration of the prefectures (cf. Carroll 1972). The location of the points representing prefectures are unaltered or the configuration of prefectures, given by the longitude and the latitude of the capital of the prefecture, is kept unchanged.

5 Results

The external analysis in the two-dimensional Euclidean space resulted in the badness-of-fit measure stress of 0.532. The derived configuration is shown in Fig. 1. Each prefecture is represented as a point and a circle centered at the point representing the prefecture.

In Fig. 1 only seven prefectures Hokkaido (#1), Saitama (#11) , Chiba (#12), Tokyo (#13), Kanagawa (#14), Fukuoka (#40), and Okinawa (#47) have their circles, and the other 40 prefectures are represented only by a point. Fukuoka has the largest radius, and Hokkaido has the second largest radius. Tokyo has the smallest radius which is zero by the normalization (Okada and Imaizumi 1987, p. 83, equations (2)–(4)). Kanagawa, Saitama, and Chiba have the second, third and fourth smallest radii. Okinawa has the fifth smallest radius. Each of the other 40 prefectures has the radius of the size which is larger than that of Okinawa and is smaller than that of Hokkaido. The reason for omitting circles of the other 40 prefectures in Fig. 1 is to prevent a too complicated configuration of prefectures. The map of Japan superimposed over the configuration of 47 prefectures was not derived from the analysis, but the map was added to facilitate the interpretation of the configuration. The derived radius is shown in Table 1.

Radii shown in Table 1 tell that prefectures surrounding Tokyo and Tokyo itself (Tokyo, Kanagawa, Saitama, and Chiba), which are in the urbanized area

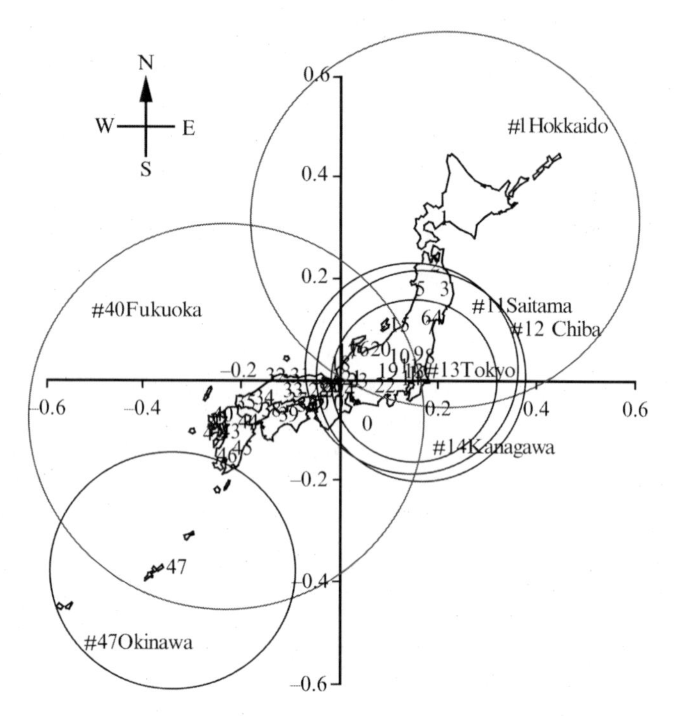

Fig. 1 Configuration of 47 prefectures derived by external asymmetric multidimensional scaling

have smaller radii, and that other urbanized areas (Aichi (#23), Kyoto (#26), and Osaka (#28)) do not have small radii. Table 1 also tells that prefectures in the rural area tends to have larger radii, while Okinawa which is in the rural area has the small radius, and that prefectures in the northeastern part of Japan tend to have the smaller radius than prefectures in the southwestern part of Japan have.

In the present analysis the larger radius means that the female in that prefecture has the larger tendency to get married to the male from the other prefectures, and that the male in that prefecture has the smaller tendency to get married to the female from the other prefectures. On the other hand, the smaller radius means that the female in that prefecture has the smaller tendency to get married to the male from the other prefectures, and that the male in that prefecture has the larger tendency to get married to the female from the other prefecture. The female of four prefectures including Tokyo (Tokyo, Kanagawa, Saitama, and Chiba) has the smaller tendency to get married to the male from the other prefectures. The female of Okinawa also has the same tendency. The female of Hokkaido and prefectures in the rural area have the larger tendency to get married to the male from the other prefectures.

Table 1 Radius derived by the external analysis

Prefecture	Radius	Prefecture	Radius
1 Hokkaido	0.382	25 Shiga	0.226
2 Aomori	0.221	26 Kyoto	0.240
3 Iwate	0.249	27 Osaka	0.247
4 Miyagi	0.216	28 Hyogo	0.220
5 Akita	0.227	29 Nara	0.220
6 Yamagata	0.212	30 Wakayama	0.228
7 Fukushima	0.230	31 Tottori	0.226
8 Ibaragi	0.214	32 Shimane	0.215
9 Tochigi	0.209	33 Okayama	0.243
10 Gunma	0.226	34 Hiroshima	0.266
11 Saitama	0.114	35 Yamaguchi	0.275
12 Chiba	0.115	36 Tokushima	0.209
13 Tokyo	0.000	37 Kagawa	0.198
14 Kanagawa	0.069	38 Ehime	0.225
15 Niigata	0.220	39 Kochi	0.231
16 Toyama	0.209	40 Fukuoka	0.398
17 Ishikawa	0.226	41 Saga	0.243
18 Fukui	0.223	42 Nagasaki	0.251
19 Yamanashi	0.214	43 Kumamoto	0.217
20 Nagano	0.218	44 Oita	0.229
21 Gifu	0.249	45 Miyazaki	0.264
22 Shizuoka	0.199	46 Kagoshima	0.265
23 Aichi	0.181	47 Okinawa	0.155
24 Mie	0.214		

6 Discussion

The relationships in the marriage among 47 Japanese prefectures were analyzed by the external one-mode two-way asymmetric multidimensional scaling. The external analysis derives the radius of each prefecture keeping the externally given configuration of prefectures unchanged. Tokyo and three neighboring prefectures have the smaller radii. Prefectures in rural areas tend to have the larger radii. The derived radius represents the outward tendency of the female of the prefecture (to get married to the male of the other prefectures) in the present analysis.

The external analysis was done in the present study. This means that the geographical relationships among prefectures are fixed and unchanged in the analysis, and that the radii representing asymmetry of the relationships in the marriage among prefectures were derived under the fixed geographic relationships of prefectures, suggesting that the effect of the geographical relationships among prefectures plays an important role in the analysis (Okada 2006a). To examine the effect of geographic relationships on the marriage among prefectures, the data were analyzed by the asymmetric cluster analysis (Okada 2000; Okada and Iwamoto 1996). The analysis using the algorithm corresponding to the complete linkage of the conventional

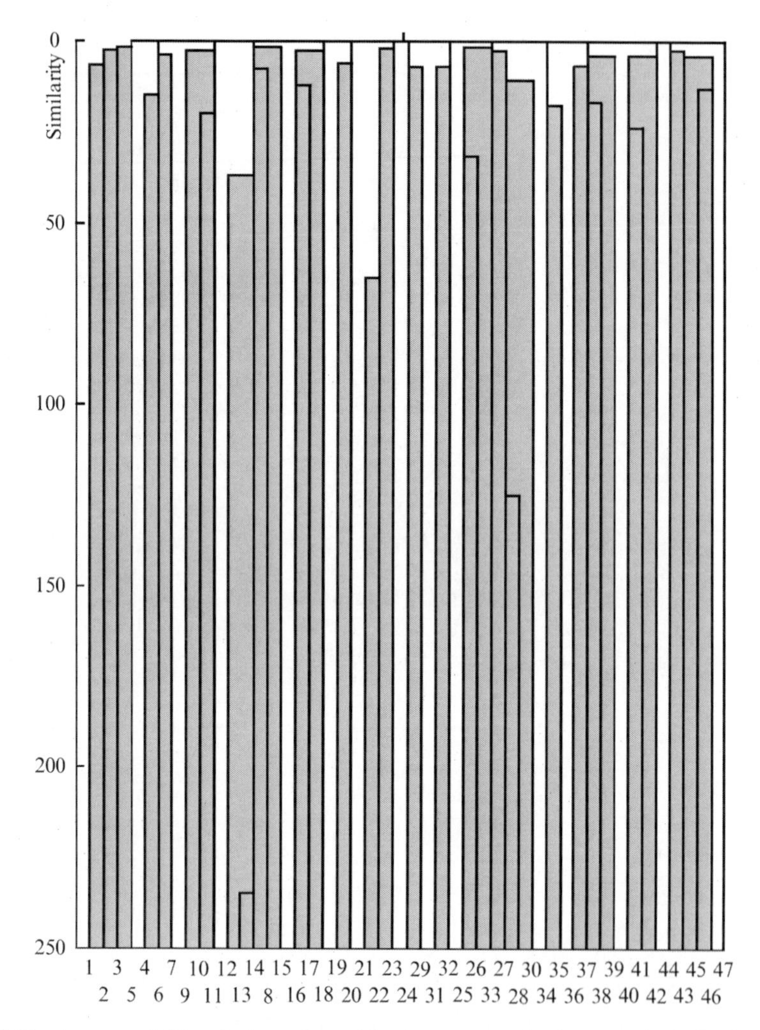

Fig. 2 Dendrogram of 47 prefectures derived by the asymmetric cluster analysis. Each number at the *bottom* represents the code representing each prefecture shown at the first and the fourth columns of Table 1. Each *shaded area* represents a cluster (Cluster 1 to 15 from left to right). Cluster 15 consists of only one prefecture, Okinawa (#47)

(symmetric) cluster analysis resulted in the dendrogram of 47 prefectures shown in Fig. 2 (Okada 2006b).

In Fig. 2, 15 clusters are gathered into one cluster at the final step of the clustering when the similarity is 0. This means that there was no marriage between any two clusters among these 15 clusters. This is why these 15 clusters form one cluster simultaneously when the similarity of 0. The 15 clusters are (from left to right of Fig. 2);

Cluster 1: 5 Akita, 1 Hokkaido, 2 Aomori, 3 Iwate
Cluster 2: 6 Yamagata, 4 Miyagi, 7 Fukushima
Cluster 3: 11 Saitama, 9 Tochigi, 10 Gunma
Cluster 4: 14 Kanagawa, 12 Chiba, 13 Tokyo, 8 Ibaragi, 15 Niigata
Cluster 5: 18 Fukui, 16 Toyama, 17 Ishikawa
Cluster 6: 20 Nagano, 19 Yamanashi
Cluster 7: 23 Aichi, 21 Gifu, 22 Shizuoka
Cluster 8: 24 Mie, 29 Nara
Cluster 9: 32 Shimane, 31 Tottori
Cluster 10: 33 Okayama, 25 Shiga, 26 Kyoto, 27 Osaka, 28 Hyogo, 30 Wakayama
Cluster 11: 34 Hiroshima, 35 Yamaguchi
Cluster 12: 37 Kagawa, 36 Tokushima, 38 Ehime, 39 Kochi
Cluster 13: 42 Nagasaki, 40 Fukuoka, 41 Saga
Cluster 14: 44 Oita, 43 Kumamoto, 45 Miyazaki, 46 Kagoshima
Cluster 15: 47 Okinawa

The first prefecture of each cluster is the center of the cluster (Okada and Iwamoto 1996). While the 15th cluster consists of only one prefecture (Okinawa), each of the other 14 clusters is formed by prefectures which neighbor on each other except for Cluster 1. In the case of Cluster 1, while Hokkaido (#1) is an island but close to Aomori (#2), the other three prefectures [Akita (#5), Aomori (#2), and Iwate (#3)] neighbor on each other. This suggests that the geographical relationships among prefectures have the significant effect on the regional relationships among prefectures on the marriage. The result of the asymmetric cluster analysis validates the external analysis of asymmetric multidimensional scaling using the geographical map of prefectures to analyze the frequency of the marriage among prefectures, because the relationships among prefectures are strongly influenced by the location of prefectures.

Two aspects should be mentioned in the present study. One is that the present data seem somewhat biased. The data were collected by a matchmaking service company. The frequency of the marriage between any two prefectures is the frequency of the marriage when both the female and the male registered themselves with the service. When either of the female or the male or both of them did not registered themselves with the service, their marriage is not included into the data (Tanaka 2005). Because the cost of registering one with the service is expensive (1,500–4,500 Euros for 2 years), only wealthy people might register themselves with the service. The other is that the frequency of cross-classified data is represented by a distance in a multidimensional Euclidean space. This can be inappropriate (DeRooij and Heiser 2003, 2005), even if the monotone transformation is used. This should be examined in the future study.

Acknowledgements The author would like to express his appreciation to two anonymous referees for their valuable reviews. The present paper was prepared, in part, when the author was at the Rikkyo (St. Paul's) University.

References

CARROLL, J.D. (1972): Individual Differences and Multidimensional Scaling. In: R.N. Shepard, A.K. Romney, and S.B. Nerlove (Eds.): *Multidimensional Scaling: Theory and Applications in the Behavioral Sciences. Volume 1 Theory*. Seminar, New York, 105–155.

DeROOIJ, M. and HEISER, W.J. (2003): A Distance Representation of the Quasi-Symmetry Model and Related Distance Models. In: H. Yanai, A. Okada, K. Shigemasu, and Y. Kano (Eds.): *New Developments in Psychometrics*. Springer, Tokyo, 487–494.

DeROOIJ, M. and HEISER, W.J. (2005): Graphical Representations and Odds Ratios in a Distance-Association Model for the Analysis of Cross-Classified Data. *Psychometrika, 70*, 1–24.

OKADA, A. (2000): An Asymmetric Cluster Analysis Study of Car Switching Data. In: W. Gaul, O. Opitz, and M. Schader (Eds.): *Data Analysis: Scientific Modeling and Practical Application*. Springer, Berlin, 495–504.

OKADA, A. (2006a): Regional Closeness in Marriage among Japanese Prefectures. In *Proceedings of the Second German-Japanese Joint Symposium on Classification.*

OKADA, A. (2006b): An Asymmetric Cluster Analysis Study of Relationships among Japanese Prefectures in Marriage. In *Proceedings of the IFCS Conference 2006.*

OKADA, A. and IMAIZUMI, T. (1987): Nonmetric Multidimensional Scaling of Asymmetric Proximities. *Behaviormetrika, 21*, 81–96.

OKADA, A. and IMAIZUMI, T. (2005): External Analysis of Two-mode Three-way Asymmetric Multidimensional Scaling. In: C. Weihs, and W. Gaul (Eds.): *Classification – The Ubiquitous Challenge*. Springer, Heidelberg, 295–288

OKADA, A. and IWAMOTO, T. (1995): An Asymmetric Cluster Analysis Study on University Enrollment Flow among Japanese Prefectures. *Sociological Theory and Methods, 10(1)*, 1–13 (in Japanese).

OKADA, A. and IWAMOTO, T. (1996): University Enrollment Flow among the Japanese Prefectures: A Comparison before and after the Joint First Stage Achievement Test by Asymmetric Cluster Analysis. *Behaviormetrika, 23*, 169–185.

SLATER, P. (1976): A Hierarchical Regionalization of Japanese Prefectures Using 1972 Interprefectural Migration Flows. *Regional Studies, 10*, 123–132.

TANAKA, T. (2005): IT Jidai no Kekkon Joho Sabisu [Matchmaking Service Using Information Technology]. *ESTRELA 134*, 27–32 (in Japanese).

Socioeconomic and Age Differences in Women's Cultural Consumption: Multidimensional Preference Analysis

M. Nakai

Abstract The differences in cultural preferences reflect the individual differences in a person's socialization process, which may differ according to people's social positions. Using 1,396 female data from a national sample in Japan in 1995, we analyzed individually different cultural preferences with multidimensional preference analysis (MDPREF). The results show that there are underlying dimensions of preference interpreted as accessibility or popularity of cultural activities and cultural prestige, which roughly correlate with the respondents' educational and occupational statuses.

1 Social Patterning of Cultural Consumption

Cultural participation reflects a person's preferences and choices. Several studies have been theorized with the notion that the cultural preferences, or "tastes" can be a marker of social class distinctions (Veblen 1899; Bourdieu 1979). For example, in his influential work, Bourdieu shows the social patterning of cultural taste. He found that family socialization process and educational experiences are the primary determinants of cultural taste.

In Japan, however, there is some controversy regarding the relationship between social stratification position and cultural taste. Some Japanese sociologists have questioned about the validity of Bourdieu's perspective in regard to Japanese society today, since in Japan quite a few men enjoy popular culture, such as karaoke and reading comic books, regardless of their educational and social status.

However, over the past few years it has been suggested that distinctive consumption practices in culture exist in Japanese society, as well as in Western societies: the hierarchical classification of cultural tastes, such as legitimate, middlebrow, and

M. Nakai
Department of Social Sciences, College of Social Sciences,
Ritsumeikan University, 56-1 Toji-in Kitamachi, Kita-ku, Kyoto 603-8577, Japan,
E-mail: mnakai@ss.ritsumei.ac.jp

A. Okada et al. (eds.), *Cooperation in Classification and Data Analysis*,
Studies in Classification, Data Analysis, and Knowledge Organizations.
DOI: 10.1007/978-3-642-00668-5_19, © Springer-Verlag Berlin Heidelberg 2009

popular, correspond to social class. Distinctive patterns in cultural consumption and early exposure to highbrow culture depend on social characteristics such as age group, educational level, gender and occupational status (Akinaga 1991; Kataoka 1992, 1998). In recent research, we also found the suggestion that criteria for the classification of cultural activities are weighted differently by women from different social subgroups (Nakai 2005). Thus it is hypothesized that the women from a different social group have a different perceptual set of cultural activities. These cultural preferences and lifestyles might maintain social boundaries (Warde 1994; Peterson and Kern 1996).

In most available research, the aspects of culture other than so-called cultural prestige have received less attention. Therefore this paper has three purposes. First, to reveal the structure of cultural preferences. Second, to find out how the dimensions of cultural preferences differ among people from different social groups. Third, to suggest a procedure that helps us to interpret differences in preferences among subjects when the number of subjects, which comprise heterogeneous social subgroups, is large.

2 Data

The data is from a nationally representative survey conducted in 1995 of social stratification and mobility in Japan. Of approximately 4,000 men and women sampled, aged 20–69, 2,653 (1,248 men and 1,405 women) were successfully interviewed. This data provides a wide range of sociodemographic characteristics and information about the cultural participation of each respondent. We selected 1,396 female respondents for our analysis to portray the structure of women's cultural preferences, as we excluded nine women who had no response to the 12 cultural activities.

We focused primarily on women because more and more women participate in the realm of volunteer work and cultural activities. The value of these nonmarket spheres of unpaid or leisure-time activities might be underestimated, but they may generate a wide range of positive social outcomes.

The following 12 cultural activities have been selected for investigation: (a) classical music performances and concerts, (b) museums and art exhibitions, (c) traditional Japanese kabuki, noh plays or Japanese puppet show (bunraku) performances, (d) karaoke, (e) pachinko, (f) golf, skiing or tennis, (g) flower arrangement, tea ceremony or calligraphy, (h) Japanese poetry (tanka or haiku), (i) community or volunteer work, (j) reading novels or books about history, (k) reading sports journals or women's weeklies, and (l) home activities such as baking bread and sweets. The ratings were made on five-point frequency scales that covered the past few years.

The data from all respondents were analyzed by using multidimensional scaling analysis of preference data, or MDPREF (Carroll 1972; Okada and Imaizumi 1994). MDPREF provides a graphical interpretation of individual preferences. In the present analysis, the frequency rating scales of cultural participation were used as input because this rating of cultural practice is believed to reflect the subjects' preferences for cultural activity. The responses were standardized to have zero mean and unit variance over all the respondents.

3 Results

The analyses in one through five dimensions gave the proportions of variance accounted for (VAF) of 0.176, 0.317, 0.429, 0.530 and 0.616, respectively. These figures of goodness of fit and the interpretability suggest that adopting a three-dimensional result (though VAF [42.9%] is not very high).

3.1 Dimensions of Cultural Preference

Cultural activities are represented by points in a dimensional space shown in Fig. 1. The first dimension represents the accessibility or popularity of cultural activities. The activities that are easily accessible and popular in general (e.g., reading novels,

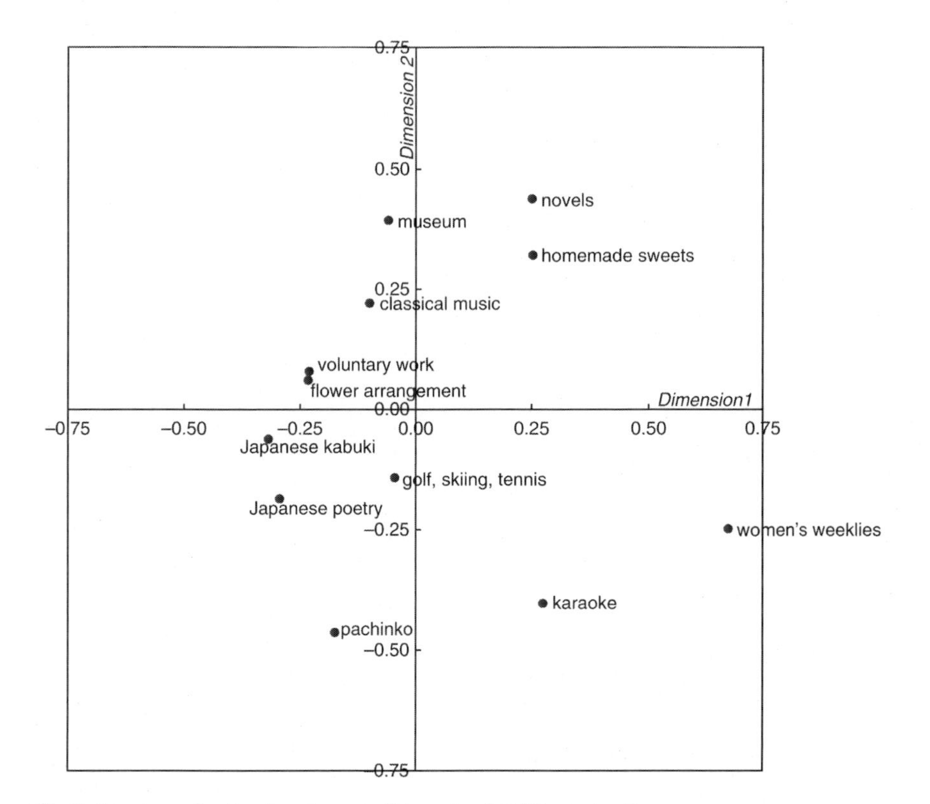

Fig. 1 Structure of cultural preference: Dimension 1 × Dimension 2

karaoke, reading women's weeklies) are in the right half, whereas those with limited following (e.g., Japanese kabuki, Japanese poetry, pachinko) are in the left half. So the horizontal dimension corresponds to the size of the consumer group and then separates universal and unpopular activities.

The second dimension is interpreted as cultural prestige. It is the criterion for consumption choices typically identified in previous studies. This vertical dimension roughly separates genteel or refined cultural activities and lowbrow activities in the upper half and lower half, respectively. Such practices as the appreciation of classical music, visiting art exhibitions and participating in community or volunteer work are evaluated positively, whereas on the other side of the dimension, a couple of cultural activities that are popular, but lowbrow (karaoke, reading women's weeklies), appear.

Next, from the configuration of stimulus points embedded in a two dimensional space in Fig. 2, dimension 3 seems to represent the types of inducement of the cultural participation, which implies cognitive motives for entree to the particular cultural sphere. The third dimension is not readily interpretable, but it roughly

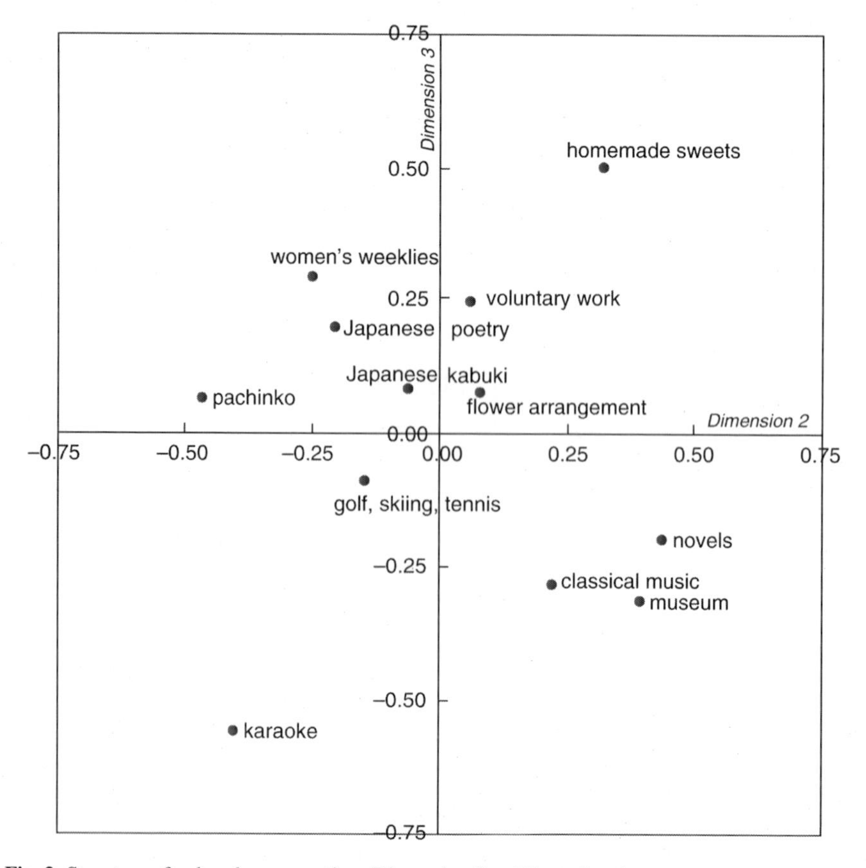

Fig. 2 Structure of cultural consumption: Dimension 2 × Dimension 3

separates the women's cultural propensity into two groups: domestic or woman-friendly culture and undomestic culture. In the upper half are activities that women participate in more than men, especially those who came through such life events as marriage or child rearing (e.g., flower arranging, volunteer work, making home-made sweets). On the other side of the dimension appear the activities through which people expand, manage and maintain relationships, especially in the world of work. People become accustomed to some of these cultures and the arts also through in-school arts and cultural education.

3.2 Differences in Cultural Preferences and Women's Social Status Groups

Next is the question of how cultural-consuming patterns are differentiated by social groups to which respondents belong. Previous research has found that the differences in cultural tastes are related to social factors such as origin, education and income. We mainly focus on respondent occupational positions and spouse occupational statuses as well as education, age and social origin. We will explore the differences in terms of the following social groups:

- Respondent's education: primary, secondary, junior college, university
- Respondent's age: 20–29, 30–39, 40–49, 50–59, 60–69
- Respondent's occupational status: professional (*prof*), managerial (*mgrl*), clerical (*cler*), sales (*sale*), skilled manual (*skil*), semi-skilled manual (*sskl*), non-skilled manual (*nskl*), agricultural (*agri*), housewife
- Husband's occupational status: professional (*H-prof*), managerial (*H-mgrl*), clerical (*H-cler*), sales (*H-sale*), skilled manual (*H-skil*), semi-skilled manual (*H-sskl*), non-skilled manual (*H-nskl*), agricultural (*H-agri*), not married (*single*)
- Father's occupational status: professional (*F-prof*), managerial (*F-mgrl*), clerical (*F-cler*), sales (*F-sale*), skilled manual (*F-skil*), semi-skilled manual (*F-sskl*), non-skilled manual (*F-nskl*), agricultural (*F-agri*)

Emboldened words in parentheses are used to represent each occupational category, and abbreviation H and F stands for the husband's and father's indicator in the figure, respectively.

It is not straightforward to interpret individual differences when the number of subjects is large. Therefore we devised the following procedure. The responses for each of the 12 cultural activities within each social group are averaged to produce individual group means. The group means are viewed as the responses of one subject. Then by utilizing the matrix representing group means, we embedded the preference for each social group in a multidimensional space of cultural activities found by MDPREF. This result gives a "point-vector" representation; the columns (stimuli, or cultural activities) as points, and the row (subjects, or social groups) as unit vectors. The vectors indicate the direction of increased preference for the

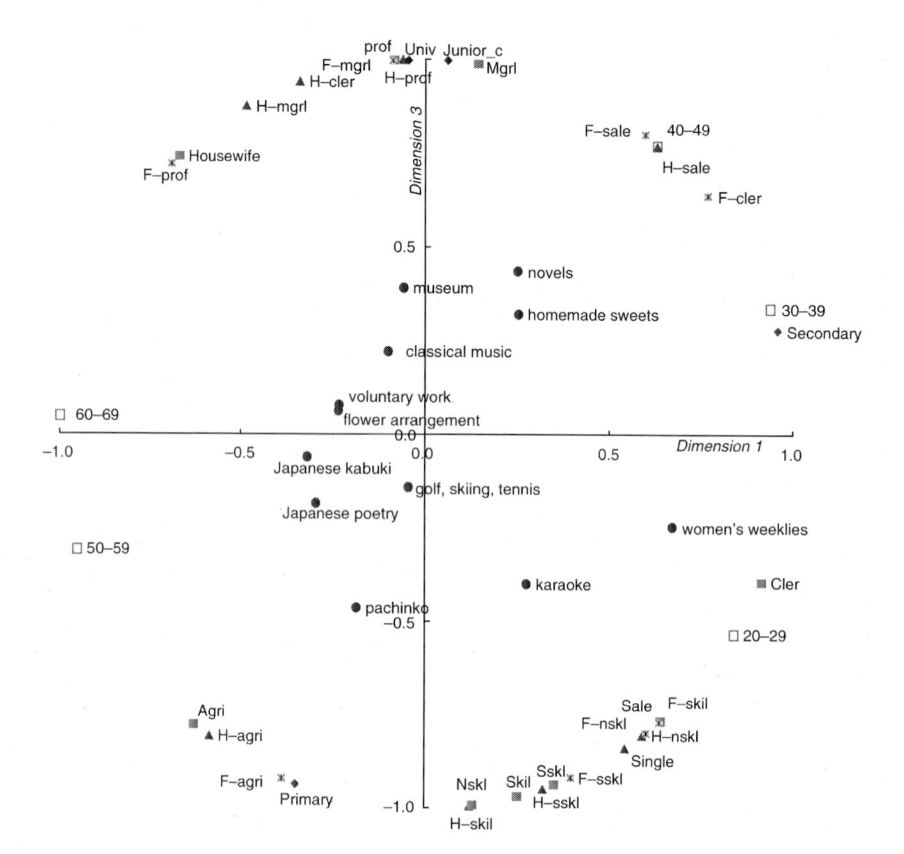

Fig. 3 Structure of cultural consumption, socioeconomic statuses and age: Dimension 1 × Dimension 2

groups. The distribution of the group vectors are shown by dots with different marks in Figs. 3 and 4.

It shows that there are numbers of vectors in the upper left and lower right near the vertical axis. As we mentioned before, the vertical dimension is interpreted as representing cultural prestige (genteel or refined vs. lowbrow or popular). Here the highly educated (university graduate) and the upper middle women (professional, managerial worker) fall into the upper quadrants close to the vertical axis and thus have a preference for highbrow or refined cultural taste, such as visiting museums and classical concerts. There are also groups in the upper left whose spouses have middle-class white-collar jobs.

On the other hand, in the lower right are numbers of groups of working-class women (skilled, semi-skilled, non-skilled manual). Therefore from the embedded vectors' relative location, these groups have opposite tastes in regard to cultural prestige. Other findings here are consistent with those of previous studies on the relationship between class and culture; that is, cultural practices and preferences are closely linked to educational level and social origin (Bourdieu 1979; Nakai 2005).

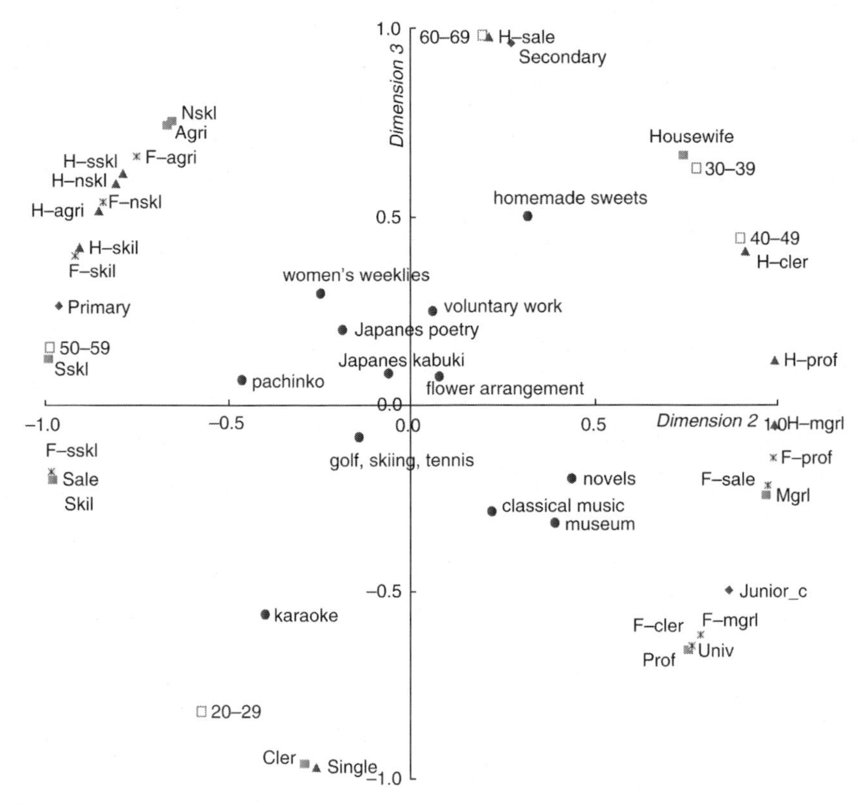

Fig. 4 Structure of cultural consumption, socioeconomic statuses and age: Dimension 1 × Dimension 3

Several group vectors are found in the lower right quadrant close to the vertical axis, but karaoke is the only cultural activity in this direction. This lack of approachable cultural activities to satisfy people's preferences indicates three things: (1) a possible niche to fill, (2) constraint of working class women on cultural participation, (3) a need to develop a better set of measurement of cultural participation.

A noticeable finding here is that women from middle-class families (both upper and lower) and whose spouses have middle-class jobs (both upper and lower) prefer highbrow culture, and those who have lower middle white-collar jobs (clerk, sales) prefer vulgar or unrefined enjoyment in general, more like the preferences of blue-collar workers.

In regard to age groups, 20s, 30s and 40s prefer universal culture, and their tastes and lifestyles change with age from popular to civilized until their 40s. In their 50s and 60s they tend to feel distaste for the prevalent activities and inactive.

Next, the configuration of second and third dimensions in Fig. 4. Vectors shown by dots of the subject groups are wide-ranging. Those who have white-collar jobs, especially single working women in their 20s, like to participate in nondomestic popular culture, which may be an opportunity to socialize. On the contrary, married middle-aged housewives hold back from participating in nondomestic culture and have a preference for womanly, domestic culture. Women in their 60s and 30s and high-school graduates tend to do so probably because they feel guilty about independent outings which may instinctively make them feel contrary to being a good mother and a good wife. This is consistent with the reports concerning social constraints that women often face; that is, women, especially married women, should give up their own spare-time pleasures (Shaw 1994). The housewives' cultural participation is influenced by their stereotypical gender role expectations. Our empirical findings suggest that the womanly or domestic cultural orientation is robust, since previous research, which analyzed the data using INDCLUS, also found a similar underlying cultural disposition (Nakai 2005).

4 Discussion and Conclusion

We analyzed 1,396 women using MDPREF and explored their cultural preferences, which vary according to their social group. From what has been discussed above, we concluded the following: (1) Three cultural criteria have been found: accessibility or popularity (being widely accepted vs. lacking general approval or acceptance); cultural prestige (highbrow vs. lowbrow or popular); and womanly or domestic taste (domestic vs. nondomestic). (2) A woman's occupational status affects her cultural preferences, as well as her educational and social backgrounds. (3) The upper middle-class woman prefers refined activities; the lower middle white-collar woman's preference is more like those of blue-collar workers. (4) The domestic vs. the nondomestic axis roughly separates housewives and working women. (5) Along with the influence of their family socialization and their statuses, women's cultural preferences are affected by spouses.

A further direction of this study will be to clarify the association of cultural participation and other social participation (in groups outside the family) from the point of view of social capital and the reproduction of social inequality.

References

AKINAGA, Y. (1991): Hierarchy of Culture and Function of Education. In: T. Miyajima and H. Fujita (Eds.): *Culture and Society.* Yushindo, Tokyo.

BOURDIEU, P. (1979): *La Distinction.* Minuit, Paris.

CARROLL, J. D. (1972): Individual Differences and Multidimensional Scaling. In: R. N. Shepard, A. K. Romney and S. B. Nerlove (Eds.): *Multidimensional Scaling: Theory and Applications in the Behavioral Sciences. Vol. 1 Theory.* Seminar, New York, 105–155.

KATAOKA, E. (1992): Social and Cultural Reproduction Process in Japan. *Sociological Theory and Methods, 11*, 33–55.

KATAOKA, E. (1998): Cultural Dimensionality and Social Stratification in Japan. *Social Stratification and Cultural Reproduction (The 1995 SSM Research Series Vol. 18)*, 87–112.

NAKAI, M. (2005): Cultural Aspects and Social Stratification of Women: Overlapping Clustering and Multidimensional Scaling. In *Proc. of the 29th Annual Conference of the German Classification Society, 246.*

OKADA, A. and IMAIZUMI, T. (1994): *Multidimensional Scaling Using Personal Computer.* Kyoritsusyuppan, Tokyo.

PETERSON, R. A. and KERN, R. (1996): Changing Highbrow Taste: From Snob to Omnivore. *American Sociological Review 61*, 900–907.

SHAW, S. (1994): Gender, Leisure, and Constraint: Towards a Framework for the Analysis of Women's Leisure. *Journal of Leisure Research 26(1)*, 8–22.

VEBLEN, T. B. (1899): *The Theory of the Leisure Class: An Economic Study in the Evolution of Institutions.* Houghton Mifflin, Boston.

WARDE, A. (1994): Consumption, Identity Formation and Uncertainty. *Sociology, 28(4)*, 877–898.

Analysis of Purchase Intentions at a Department Store by Three-Way Distance Model

A. Nakayama

Abstract This study focused on analyzing the reasons for purchase of ladies' goods. The analysis used the three-way distance model. The model explains three-way distances as the subtraction of the smallest squared distance among the three squared distances from the sum of these squared distances. The formulation is used to illustrate the idea that relationships with many differences carry more information than relationships with few differences.

1 Introduction

The purpose of the present study is to show a framework of purchase intentions at a department store. The analysis used the three-way distance model which was proposed by Nakayama (2005). The Shepard–Kruskal approach to non-metric multidimensional scaling (Kruskal 1964a,b; Shepard 1962a,b) is modified in Nakayama (2005)'s model. The Shepard–Kruskal is a simple model that is easily generalizable from a two-way distance model into a three-way distance model. Therefore, the allows the visualization of relationships among objects and makes them easier to understand. For other methods on one-mode three-way MDS, see, for example, Cox et al. (1991), De Rooij and Gower (2003), Hayashi (1989), Heiser and Bennani (1997), and Joly and Le Calvé (1995). Finally, this study compares the solution provided by Nakayama (2005)'s model with representations obtained using a one-mode two-way MDS. It then examined whether Nakayama (2005)'s model provided solutions different from those of one-mode two-way MDS. These comparison revealed new important insights, i.e., consumers follow regular patterns when making purchases.

A. Nakayama
Faculty of Economics, Nagasaki University, 4-2-1 Katafuchi, Nagasaki, Japan 850-8506,
E-mail: atsuho@nagasaki-u.ac.jp

A. Okada et al. (eds.), *Cooperation in Classification and Data Analysis*,
Studies in Classification, Data Analysis, and Knowledge Organizations.
DOI: 10.1007/978-3-642-00668-5_20,

The marketplace contains many shops and products from which consumers can choose, and merchants face strong competition when attempting to attract consumers (Woolf 1996). Consequently, a careful analysis is needed on how and when joint purchase decisions may or may not take place. Selection of an item from one category may lead to a joint purchase from another category based on the complementary nature and similar purchase cycles between the two categories. However, a joint purchase may not occur because of monetary reasons, temporal restrictions, or other observed factors. Consumer purchasing behavior has been of increasing interest and has consequently spurred many studies on how consumers react to a wide variety of marketing promotions (e.g., Guadagni and Little 1983; Gupta 1988). Therefore, this study analyzed customer survey data provided by a department store. The customer survey was intended to reveal how customers plan purchases. This study focused on socioeconomic setting, personal setting, product setting, and in-store marketing in a department store. Sheth (1983) summarized patronage behavior theory, which covers purchase behavior with respect to a specific product or service from an outlet and explains these using a vector of four behavioral outcomes, i.e., planned purchase, unplanned purchase, forgone purchase, and no purchase. Sheth (1983) described patronage behavior as a function of preference-behavior discrepancies caused by four types of unexpected events that have either no effect or act as inducements: socioeconomic setting, personal setting, product setting, and in-store marketing. "Socioeconomic setting" refers to macroeconomic conditions such as inflation, unemployment, and interest rates, as well as to social situations such as the presence of friends or relatives at the time of shopping. "Personal setting" relates to the time, money, and physical effort spent by an individual shopper at the time of shopping. "Product setting" refers to the marketing mix of the store's product class, such as brand availability, relative price structure, unexpected sales promotion, and shelf locations of various product options. "In-store marketing" relates to unexpected in-store changes such as a new brand, a change in the location of existing brands, in-store promotions, and selective sales efforts made by a salesclerk.

2 Data

The customer survey was intended to reveal how customers plan purchases. It was presented to 500 randomly selected customers, and 269 questionnaires were returned. Of these, 186 were complete and were used as the sample. The questionnaire included 15 possible responses. First, consumers were asked to choose from five motivations for buying cosmetics. Table 1 presents the results. The word(s) in italics in (Table 1) are used in Figs. 1 and 2 to refer to each motivation. Consumers were asked the same questions about ladies' wears and ladies' shoes. A $15 \times 15 \times 15$ one-mode three-way symmetric matrix was calculated from the data. The matrix indicates the frequency with which motivation from each of three motivations was emphasized simultaneously. The matrix was subsequently analyzed using Nakayama (2005)'s model.

Table 1 Reasons for purchase of ladies' goods and number of responses

	Cosmetics	Ladies' wears	Ladies' shoes
1. I wanted to buy *new items*	107	189	125
2. I had an *event* to attend such as a ceremony or one involving travel	43	180	141
3. I saw advertisement *posters* and leaflets	63	122	78
4. I received *direct mail* advertisements	95	124	53
5. The *rewards* such as discounts were great	92	194	148

3 Analysis

Three-way dissimilarities δ_{ijk} are known as "three-way dissimilarities functions," when a set of n objects is given and the indices i, j, and k represent any three objects. Furthermore, these are generalized forms of two-way dissimilarities, where δ_{ijk} represents the dissimilarity of the three categories. One configuration, X of n points $x_i = (x_{i1}, x_{i2}, \ldots, x_{ip}$, is assumed, for $i = 1, \ldots, n$, in a p-dimensional Euclidean space, where an x_i-coordinate corresponds to the point for object i. The distance among three objects, d_{ijk}, is meant to reflect the values of the dissimilarities, δ_{ijk}. Nakayama (2005) proposed a model that employs different weights for each of the three dyadic distances. The model subtracts the shortest squared distance of the three squared distances from the sum of the squared distances. This particular formulation allows isolation of the triadic relationships underlying the data; i.e., the new three-way distance d_{ijk} can be expressed as

$$d_{ijk} = \sqrt{d_{ij}^2 + d_{jk}^2 + d_{ik}^2 - \min(d_{ij}^2, d_{jk}^2, d_{ik}^2)}. \tag{1}$$

In this context, the distance between two objects is assigned a heavier weight if the two objects differ greatly. The smallest dissimilarity between two objects is assigned a weight of 0, and this distance is excluded from the summation. This formulation is based on the theory that relationships with many differences supply more information than relationships with few differences. As a result, one three-way distance based on three almost-equal dyadic distances would be smaller than another three-way distance based on one short dyadic distance and two long dyadic distances, when the sum of the three distances between each pair of objects is equal. Note that the present study has modified the Shepard–Kruskal approach to nonmetric multidimensional scaling (Kruskal 1964a, b; Shepard 1962a, b). The Shepard–Kruskal model is simple and can be easily generalized from a two-way distance model into a three-way distance model. Therefore, the model proposed here allows the visualization of relationships among objects, making them easier to understand.

Equation (1) is a three-way distance function if and only if:

$$d_{ijk} \geq 0, \tag{2}$$

$$d_{ijk} = d_{ikj} =, \ldots, d_{(\text{all permutations})}, \tag{3}$$

$$d_{ijk} = 0 \text{ only if } i = j = k, \tag{4}$$

$$d_{iji} = d_{iji}, \tag{5}$$

$$2d_{ijk} \leq d_{ikl} + d_{jkl} + d_{ijl}. \tag{6}$$

Therefore, three-way distances d_{ijk} must satisfy non-negativity and three-way symmetry, or all i, j, and k. The third condition above stipulates that three-way self-dissimilarities should not differ from 0, while the fourth specifies that, when one object is identical to another, the lack of resemblance between the two nonidentical objects should remain invariant, regardless of which two are identical. By symmetry, this condition must also be satisfied for $d_{iij} = d_{jji}, d_{iji} = d_{jij}, d_{ijj} = d_{jji}$, and so on. The last condition plays a role similar to that of the triangle inequality in the context of two-way distances. In summary, (1) will be called a "three-way distance," if and only if (1) satisfies (2)–(6).

The Shepard–Kruskal model for non-metric MDS is applied to the Nakayama (2005)'s model. In order to obtain the best-fitting configuration, the initial configuration is first determined, and then the three-way distances d_{ijk} are calculated. Next, a monotone regression is used to find the that $\hat{d}_{ijk}$ satisfy the following conditions:

$$\delta_{ijk} < \delta_{rst} \Rightarrow \hat{d}_{ijk} < \hat{d}_{rst} \quad \text{for all} \quad i < j < k, \; r < s < t. \tag{7}$$

The measure of badness-of-fit of d_{ijk} to δ_{ijk} is called the "stress S" and is based on the stress formula defined below (Kruskal and Carroll 1969)

$$S = \sqrt{\sum_{i<j<k}^{n} (d_{ijk} - \hat{d}_{ijk})^2 \Big/ \sum_{i<j<k}^{n} (d_{ijk} - \bar{d}_{ijk})^2}. \tag{8}$$

The analysis used the maximum dimensionalities of nine through five. Therefore, the first stress values were obtained in nine- through unidimensional spaces, the second stress values were obtained in eight- through unidimensional spaces, the third stress values in seven- through unidimensional spaces, etc. The smallest stress value in each dimensional space was selected as the minimum stress value for that dimensional space. The resulting minimum stress values in five- through unidimensional spaces were 0.268, 0.279, 0.312, 0.349, and 0.395, respectively. Examination of these five stress values revealed that a two-dimensional configuration provided the most appropriate solution, and was thus the best solution: a two-dimensional configuration is easy to visualize, and relationships among the motivations are readily apparent.

4 Results

Analysis results clearly identified differences among groups of motivations and revealed the relationships among the intentions of consumers to purchase cosmetics, ladies' wears, and ladies' shoes. Motivations located in the central part of the

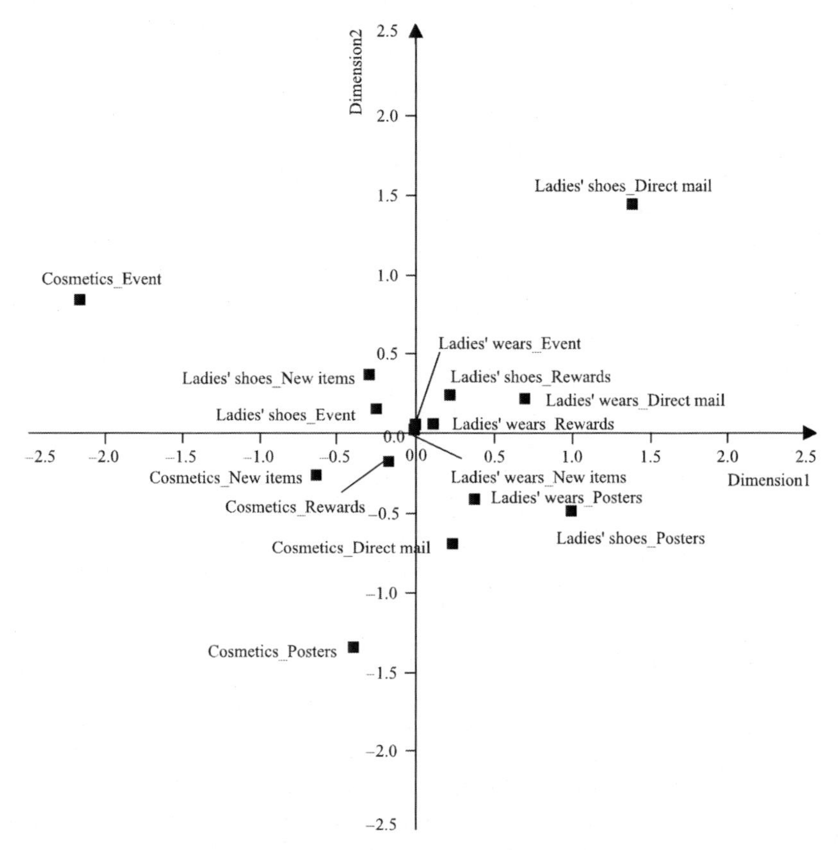

Fig. 1 The two-dimensional configuration obtained from the 15 × 15 × 15 one-mode three-way symmetric matrix by using the Nakayama (2005)'s model

configuration were often considered in conjunction with other motivations, while the position of motivations located near the edge of Fig. 1 indicates that they were considered separately. The results of the analysis clearly identified differences among groups of cosmetics, ladies' wears, and ladies' shoes. Results for cosmetics are located in the left half, ladies' wears in the lower right, and ladies' shoes in the upper right. The analysis showed that consumers' intensions were changed by each of three items.

The analysis also revealed several popular combinations of three motivations for buying. Three-motivation joint considerations occurred most frequently among ladies' wears for new items, an event, and rewards. Second in frequency was the combination of a joint consideration of two motivations from ladies' wears for new items, an event, or rewards, along with another motivation located in close proximity to these. Some examples include one motivation from ladies' shoes for an event and rewards, cosmetics for rewards, or ladies' wears due to direct mail. The third most frequent case involved any combination other than ladies' wears for new items, an event, or rewards. These include a combination of ladies' shoes for an event, rewards, and new items, ladies' shoes for an event and cosmetics for rewards and for new items, or cosmetics for rewards and ladies' shoes for an event and rewards. There exists that the finer reasons for buying over the category of three items. The analysis showed that consumers' intensions were based on certain frameworks. Therefore, strategic approaches that focus on how socioeconomic setting, personal setting, product setting, and in-store marketing influence purchases could be of major importance. Consumers' purchase intensions were generally based on each of three items of cosmetics, ladies' wears, and ladies' shoes. Strategic approaches that were planed by each three items could be appropriate for the global viewpoint. Then, several specific combinations of three motivations for buying were shown. Strategic approaches that were based on specific combinations could be suitable for the local viewpoint.

5 Discussion

The above analysis was carried out to apply the information contained in customer survey data to the expression of the Nakayama (2005)'s model. As noted above, the model seems to have revealed the relationships among consumer's intentions to purchase clearly. However, the solution had to be compared with a representation obtained from a one-mode, two-way MDS (Kruskal 1964a, b). A 15×15 one-mode, two-way symmetric matrix was calculated from the purchase data. The matrix expresses the frequency of joint purchases and was analyzed using one-mode, two-way MDS. The configurations were derived in nine- through unidimensional spaces in the same way as in the above analysis. The two- dimensional configuration obtained from the one-mode, two-way model is illustrated in Fig. 2. Figure 2 shows how Procrustes analysis was used to match the configuration from one-mode two-way model to the configuration for the Nakayama (2005)'s model. It reveals that the generalized Euclidean distance model provided a spatial representation generally similar to that from the variable weights model. Motivations located in the central part of Fig. 2 were often considered in conjunction with other motivations, while motivations located in the edge of Fig. 2 were considered separately. However, the one-mode two-way model's configuration is more difficult to interpret than that of the Nakayama (2005)'s model. A comparison of Fig. 1 to Fig. 2 reveals that some motivations in the one-mode two-way model's configuration differ in position from that of the Nakayama (2005)'s model. Some categories in the configuration of the one-mode, two-way model case are located nearer to the center of the configuration than those in the configuration of the Nakayama (2005)'s model case. These include ladies' shoes due to rewards and an event, and purchases of cosmetics for new items. The Nakayama (2005)'s model represents the triadic relationships among such motivations more accurately than the one-mode two-way model. These results indicate that the Nakayama (2005)'s model produces finer representations among three motivations than the one-mode two-way model. The solution obtained

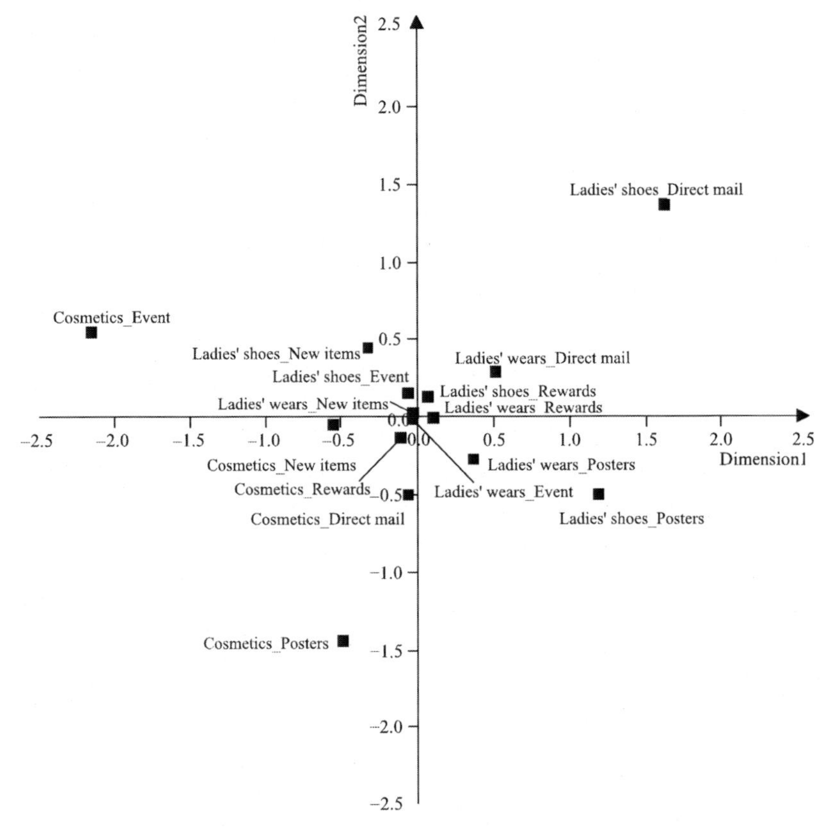

Fig. 2 The configuration for one-mode two-way MDS matched to the configuration for the Nakayama (2005)'s model by using Procrustes analysis

using the Nakayama (2005)'s model provided more information about relationships among motivations than the one-mode two-way model.

This study showed that the relationships among consumer's intentions to purchase cosmetics, ladies' wears, and ladies' shoes. The results of the Nakayama (2005)'s model clearly identified differences among groups of three items. Nakayama (2005)'s model provided a better fit for the data than did the one-mode two-way MDS. Some motivations in the Nakayama (2005)'s model configuration differ in position from that of the one-mode two-way model. However, Nakayama (2005)'s model generally provides a similar spatial representation to that of one-mode two-way MDS. Some motivations in the configuration of the one-mode two-way MDS are located nearer to the center of the configuration than those in the configuration of three-way distance model.

References

COX, T. F., COX, M. A. A., and BRANCO, J. A. (1991): Multidimensional Scaling for *n*-Tuples. *British Journal of Mathematical and Statistical Psychology, 44*, 195–206.

De ROOIJ, M., and GOWER, J. C. (2003): The Geometry of Triadic Distances. *Journal of Classification, 20*, 181–220.

GUADAGNI, P. M., and LITTLE, J. D. C. (1983): A Logit Model of Brand Choice Calibrated on Scanner Data. *Marketing Science, 2*, 203–238.

GUPTA, S. (1988): Impact of Sales Promotion on When, What and How Much to Buy. *Journal of Marketing Research, 25*, 342–355.

HAYASHI, C. (1989): Multiway Data Matrix and Method of Quantification of Qualitative Data as a Strategy of Data Analysis. In: R. Coppi and S. Bolasco (Eds.): *Multiway Data Analysis.* North Holland, Amsterdam, 131–142.

HEISER, W. J., and BENNANI, M. (1997): Triadic Distance Models: Axiomatization and Least Squares Representation. *Journal of Mathematical Psychology, 41*, 189–206.

JOLY, S., and LE CALVÉ, G. (1995): Three-Way Distances. *Journal of Classification, 12*, 191–205.

KRUSKAL, J. B. (1964a): Multidimensional Scaling by Optimizing Goodness of Fit to a Nonmetric Hypothesis. *Psychometrika, 29*, 1–27.

KRUSKAL, J. B. (1964b): Nonmetric Multidimensional Scaling: A Numerical Method. *Psychometrika, 29*, 115–129.

KRUSKAL, J. B., and CARROLL, J. D. (1969): Geometrical Models and Badness-of-Fit Functions. In: P. R. Krishnaiah (Ed.): *Multivariate analysis, Vol. 2.* Academic, New York, 639–671.

NAKAYAMA, A. (2005): A Multidimensional Scaling Model for Three-Way Data Analysis. *Behaviormetrika, 32*, 95–110.

SHEPARD, R. N. (1962a): The Analysis of Proximities: Multidimensional Scaling With an Unknown Distance Function. *Psychometrika, 27*, 125–140.

SHEPARD, R. N. (1962b): The Analysis of Proximities: Multidimensional Scaling with an Unknown Distance Function. *Psychometrika, 27*, 219–246.

SHETH, S. N. (1983): An Integrative Theory of Patronage Preference and Behavior. In: W. R. Darden and R. F. Lusch (Eds.): *Patronage behavior and retail management.* North Holland, Amsterdam, 25–28.

WOOLF, B. P. (1996): *Customer Specific Marketing: The New Power in Retailing.* Teal Books, Greenville, SC.

Facet Analysis of the AsiaBarometer Survey: Well-being, Trust and Political Attitudes

K. Manabe

Abstract The purpose of this paper is to illustrate the utility of Facet Analysis developed by Louis Guttman and his group by using the examples of the AsiaBarometer survey. The AsiaBarometer is a large scale multi-national questionnaire survey conducted at regular intervals (every year from 2003) within the Asia region, and focuses primarily on the everyday lives of ordinary people. It also asks questions about social values and norms, and political attitudes and behaviors.

1 Introduction

The third AsiaBarometer[1] Survey was conducted in the following 14 countries in south and central Asia in 2005: Afghanistan, Bangladesh, Bhutan, India, Kazakhstan, Kyrgyzstan, Maldives, Mongolia, Nepal, Pakistan, Sri Lanka, Tajikistan, Turkmenistan, and Uzbekistan. This article examines the data from three countries in each of the two regions surveyed: India, Pakistan, Sri Lanka, Kazakhstan, Mongolia, and Uzbekistan. The following must be noted regarding the type of data analysis that will be conducted using the data from these six countries. This article uses a technique of Facet Analysis developed by L. Guttman, that is, Smallest Space Analysis (SSA-I). SSA is a type of multidimensional scaling method, and illustrates graphically the structure of relationship between n items shown in a correlation matrix by the size of the distance between n points in an m-dimensional ($m < n$) space. The higher the correlation between two variables become, the smaller the distance between them on the map, and the lower the correlation, the larger the distance. Usually a two-dimensional (plane) or three-dimensional (cube) space is used to visually depict the relationship between items

[1] AsiaBarometer official website: http://www.asiabarometer.org.

K. Manabe
School of Sociology, Kwansei Gakuin University, 1-1-155 Uegahara, Nishinomiya 662-0811, Japan, E-mail: kazufumi.manabe@nifty.com

A. Okada et al. (eds.), *Cooperation in Classification and Data Analysis*,

Studies in Classification, Data Analysis, and Knowledge Organizations.
DOI: 10.1007/978-3-642-00668-5_21,

as Levy (1994) and Manabe (2001). SSA is likely to be applicable as a very effective tool in examining equivalence of measurement, a major concern when conducting cross-national surveys. In a cross-national survey that uses the same wording for the same question items can yield an SSA map showing the same spatial structure (constellation), it is highly likely that "commonality of meaning" can likewise be established in those countries. Thus, it is highly likely that the equivalence of measurement can be ensured. This is an important reason for using SSA.

2 Results of Data Analysis

2.1 Well-being

The group of question items classified under the heading of "well-being" contains 18 question items: Q5 (a question item on the feeling of happiness), Q6a-p (16 question items on life satisfaction), and Q7 (a question item on the objective standard of living). Here I created a correlation matrix showing the relationships between question items on well-being in the six countries of India, Kazakhstan, Mongolia, Pakistan, Sri Lanka, and Uzbekistan, and obtained the following six two-dimensional maps using SSA. The circles shown in these maps reflect an effort to apply meaning to (interpret) the spatial distribution of the question items based on the empirical law of L. Guttman's Facet Theory. To make those interpretations easier to decipher, the number of each question item is enclosed in either a ○ (circle) or △ (triangle). A circle indicates that the correlation coefficient with the "feeling of happiness" (Q5) is 0.3 or higher, while a triangle indicates that it is 0.2 or higher (Fig. 1).

The SSA map shows that the structure of well-being is depicted by a series of concentric circles with Q5, the feeling of happiness, at the center. All the questions are then distributed throughout these circles, at various levels in accordance with their content relevance with Q5. The numbers of the question items closest to the "feeling of happiness" are enclosed in a circle, while the next-closest are enclosed in a triangle. The numbers of question items located outside the space are not enclosed in either a circle or triangle. This kind of analysis makes it possible to visually discern how close a relationship there is in each country between the people's "feeling of happiness" and their "life satisfaction"? in other words, the aspects of the relationship between the "feeling of happiness" and "life satisfaction." I would now like to describe the unique aspects of the SSA maps for each of the six countries examined. First, however, it may prove useful to identify a commonality among all of the countries, that is, the position of the four question items regarding public safety, the environment, social welfare, and the democratic system. Among the question items asked regarding life satisfaction, these four were slightly different in terms of their content than the others. While the other question items tended to address the "personal sphere," these four tended more to address the "institutional sphere." In a sense, the difference between these may correspond to the conceptual distinction

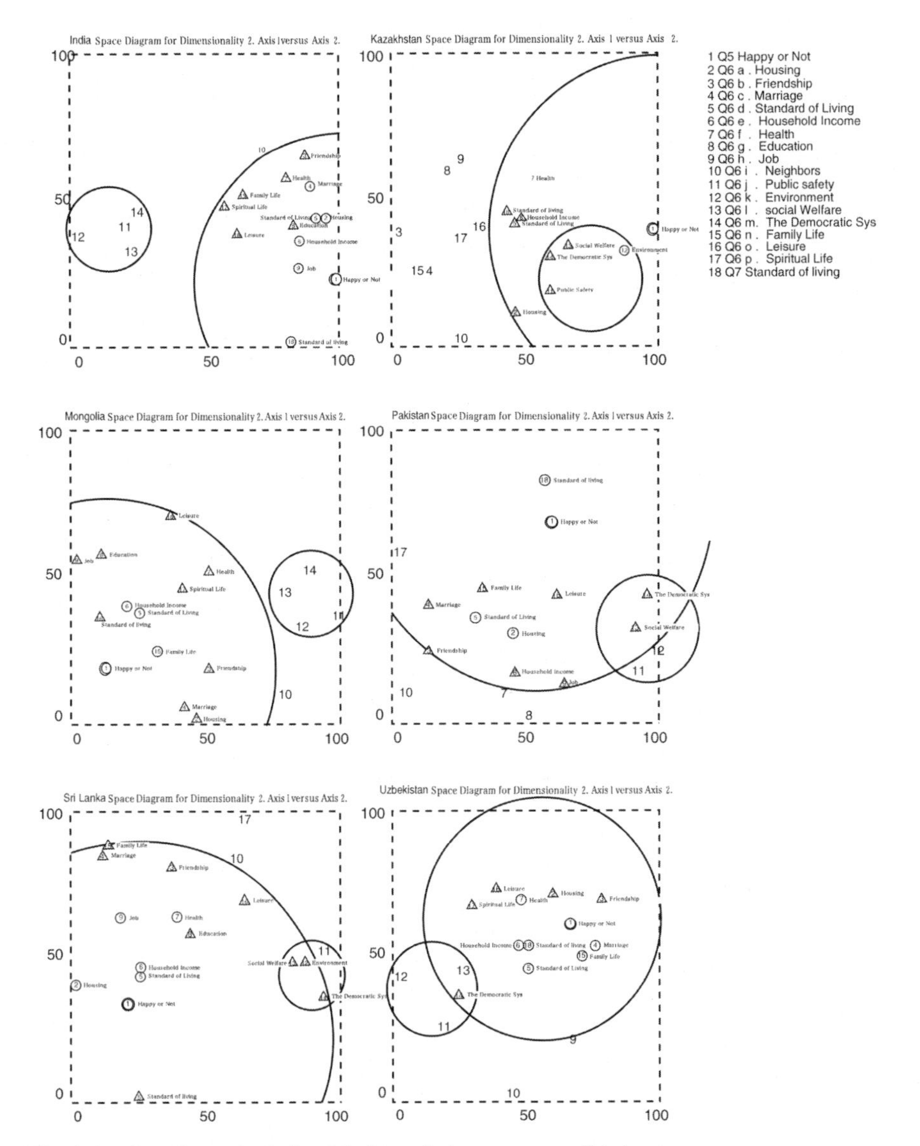

Fig. 1 Smallest Space Analysis of the interrelations among well-being items

between "small happiness" and "great happiness" proposed by Aoki (2003). Turning our attention back to the SSA maps for these questions items, we find, based on an investigation of these meanings, that these four items comprise a single independent space in all of the countries. This can therefore be identified as a common facet seen in every country. In some cases, the space composed of these four question items is located far away from the space comprised of the other question items, while in others it overlaps with part or all of the space comprised of the other question items. Examples of the former are India and Mongolia, while examples of the latter are Kazakhstan (fully overlapping), Sri Lanka (three question items overlapping), Pakistan (two question items overlapping) and Uzbekistan (one question item overlapping). Even in those cases exhibiting the most overlap, the numbers of the overlapping question items – except in the question item on the environment in Kazakhstan – were at most only enclosed in a triangle, indicating that even the degree of overlap was not all that large. While on the one hand this clarifies the differences between countries, on the other, it also allows generalizable propositions that go beyond nation specific differences – such as the cross-cultural conceptualization of "small happiness" and "great happiness"? to be established. To determine where these nation specific differences come from would require an investigation of the various economic, political, social, and cultural conditions in each country. Next, there are two types of question items that have been characterized as belonging to the "personal sphere": instrumental question items, on such topics as housing, income, standard of living, job, health, and education, and consummatory question items, on such topics as marriage, family life, friends, neighbors, and leisure. Based on this distinction, the SSA maps in each country show that there are countries in which people's "feeling of happiness" is close to the instrumental question items (such as India and Sri Lanka) and countries in which it is close to both the instrumental and the consummatory question items(such as Uzbekistan and Mongolia). Explaining this distinction will require individual intensive analyses of the relationships between the social and cultural factors in each country.

2.2 Trust

The battery of 22 question items addressed in this analysis are divided into three question items on "interpersonal trust" (Q11, Q10, and Q12) and 19 question items on "institutional trust" (Q27a-s). The former three question items are as follows: "Generally, do you think people can be trusted or do you think that you can't be too careful in dealing with people?" (Q10), "Do you think that people generally try to be helpful or do you think that they mostly look out for themselves?" (Q11), and "If you saw somebody on the street looking lost, would you stop to help?" (Q12). The latter 19 question items ask if the respondent trusts institutions, organizations and systems, such as "the central government," "your local government," "the army," "the legal system," "the police," and "parliament." Here again, I have created a correlation matrix showing the relationships between these question items for the six countries being examined, and have obtained the following six two-dimensional maps using SSA (Fig. 2).

In this case, the similarities shared by the countries are even more remarkable than the differences between them. The question items related to people's social trust are shown in concentric circles in all of the countries, but question items 1, 2, and 3 on interpersonal trust are located in the inner concentric circle, while question items 4–22 on institutional trust are positioned in the outer concentric circle. In other

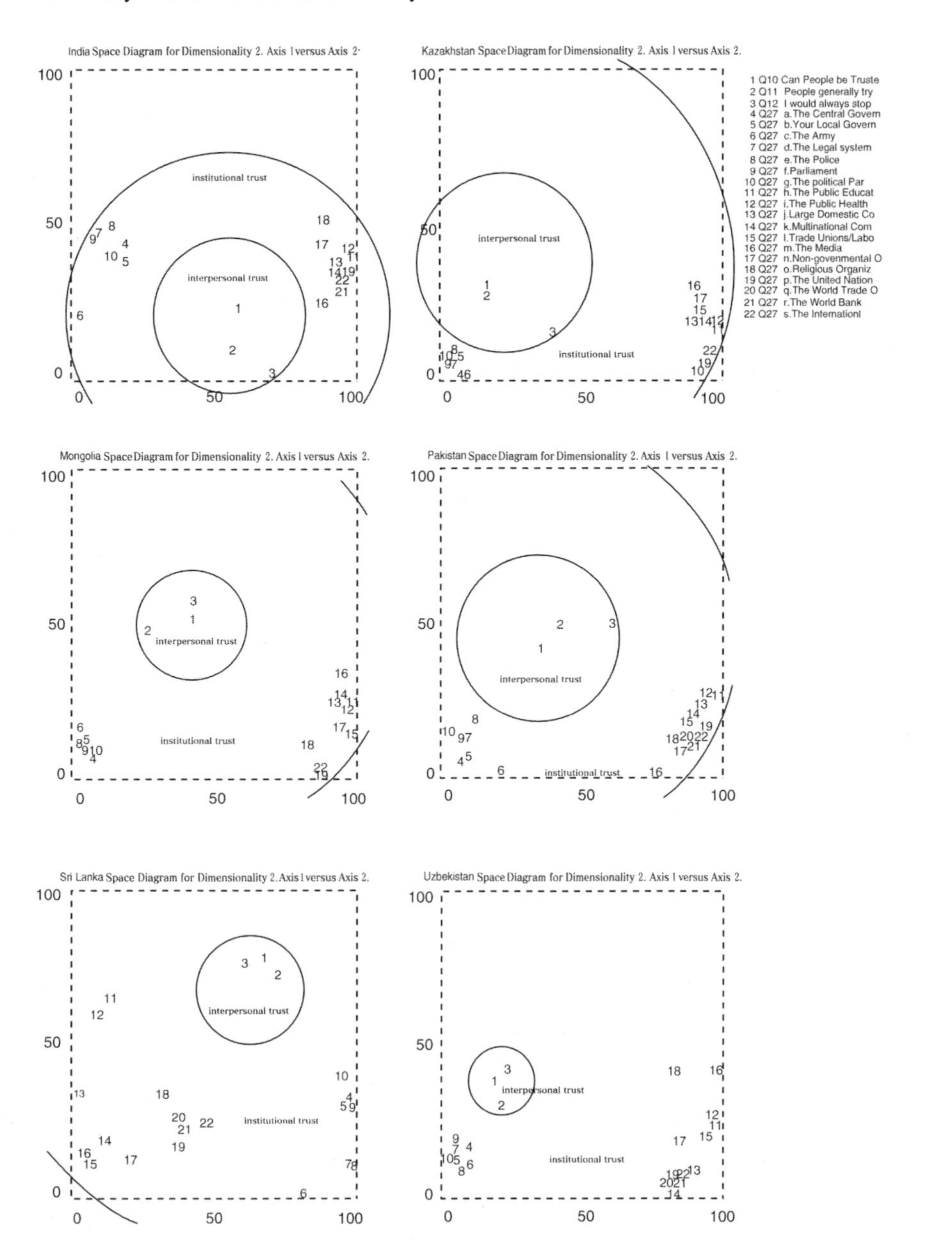

Fig. 2 Smallest Space Analysis of the interrelations among trust items

words, there tends to be a disjuncture between interpersonal trust and institutional trust. Specifically, the tendency to say "I trust people," but "I don't trust institutions" appears to be consistent across all of the countries. Even the question items labeled as "institutional trust" question items can be divided into two groups: the group from the central government to the political party, and the group from the public education system to the International Monetary Fund. The former are institutions in the "political sphere," while the latter are part of the "economic, social, and cultural sphere." This is not to say that there are no cross-national differences. While the disjuncture between the two types of trust is common across all countries, the size of that disjuncture varies. A closer inspection of the SSA maps of each of the countries shows that the positions of each of the question items are slightly different. Here, again, the maps serve two purposes. On the one hand, they facilitate the establishment of generalizable propositions – the disjuncture between interpersonal trust and institutional trust –, while on the other they serve to elucidate the specific characteristics of each of the countries. Active debates around the theme of "trust" addressed in this section have been developed over the past decade. Fukuyama (1995), one of the instigators of those debates, makes an analytical argument in his book *Trust* regarding the relationship between the social structure and the people's sense of trust in various countries. The generalizable proposition suggested above is viewed as contributing to the development of Fukuyama's argument.

2.3 Political Attitudes

The AsiaBarometer survey contains several question items intended to ascertain people's political attitudes. For the purpose of this data analysis, I will examine a battery of seven question items, Q31a-g. These question items have been used in various political attitude surveys conducted since Almond and Verba (1963). Once again, I will begin by examining the content of these question items. A careful examination reveals that these seven question items can be divided into three groups: (1) a question item regarding political duty(①), (2) question items regarding political cynicism (② ⑥ ⑦), and (3) question items regarding political efficacy (③ ④ ⑤). The purpose of this data analysis is to verify whether this conceptualization is applicable across cultures. Thus, I created a correlation matrix showing the correlations between these seven questions, and obtained the following six maps using SSA (Fig. 3).

These SSA maps show that the seven question items regarding political attitudes can be divided into three groups in virtually every country. In Pakistan, however, ⑦ Q31g ("Government officials pay little attention to what citizens like me think") belongs both in the sphere of "political cynicism" and "political efficacy." The analysis above suggests that the three-pronged conceptualization of political attitudes as political duty, political cynicism, or political efficacy is applicable across countries. In this case, too, however, further investigation of the specific conditions in Pakistan must be conducted to determine why ⑦Q31g has characteristics of both "political cynicism" and "political efficacy."

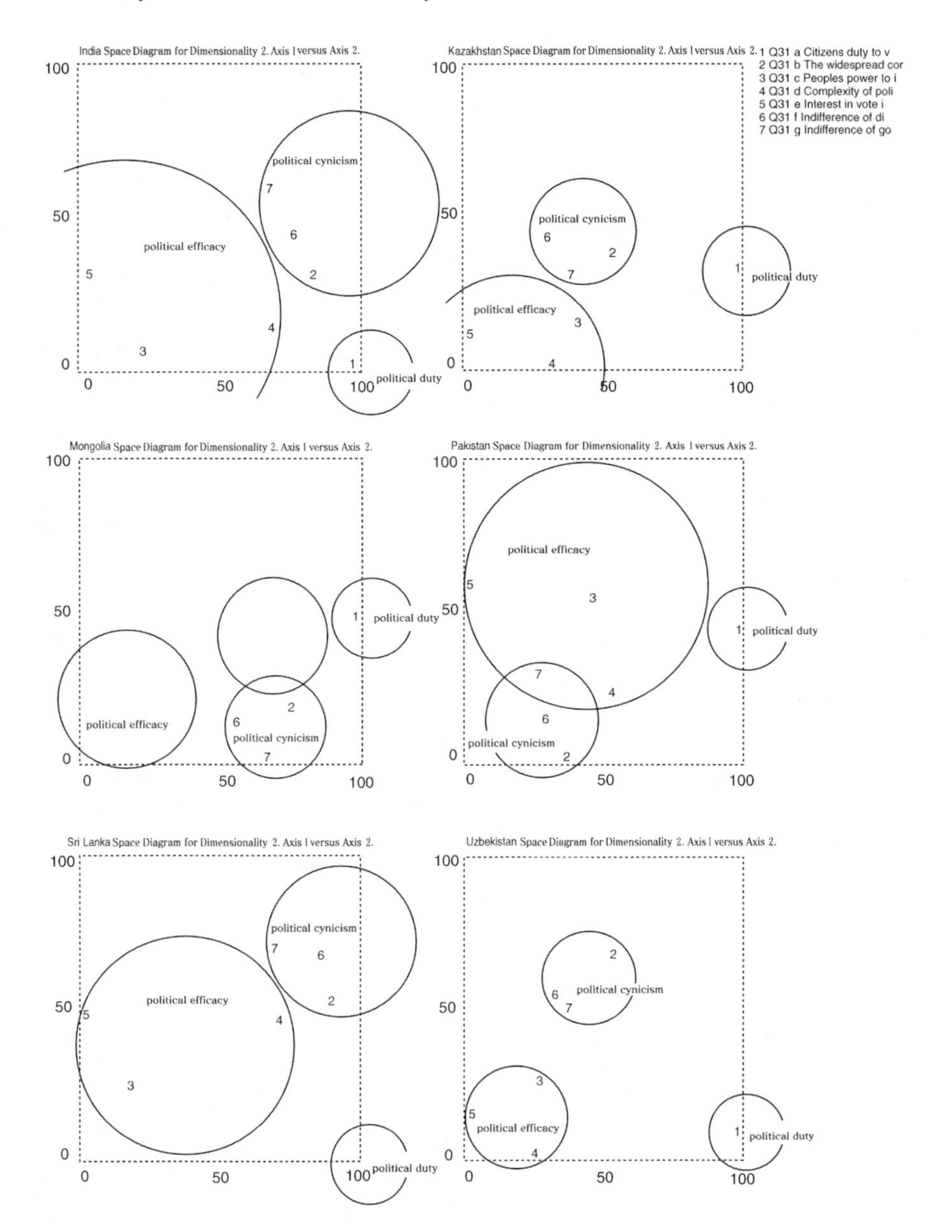

Fig. 3 Smallest Space Analysis of the interrelations among political attitudes items

3 Conclusion

By focusing on the methodological advantages of Facet Analysis, this paper has made several attempts at exploratory data analysis for the purpose of pointing out the significance of the survey, suggesting possible directions for data analysis, and establishing a starting point for future data analysis efforts. It may also be considered an attempt to perform the intellectual exercise known as "dimensional confirmation" as a first step in empirical research on social phenomena. This kind of intellectual exercise will make it possible to proceed with an analysis of relationships between multiple dimensions confirmed. These kinds of data analysis procedures constitute an approach based on what is called "data science." However, this paper also makes a suggestion about data analysis attempts from other perspectives, that is, that the AsiaBarometer Survey be used to examine hypotheses that have been deductively derived from theories developed in different disciplines, such as political science, sociology, and psychology. The AsiaBarometer Survey has a great deal of latent potential in this regard. Making the data available to the public is likely to encourage researchers in a wide variety of fields to use that data in examining their various theoretical hypotheses.

References

ALMOMD, G. A. and VERBA, S. (1963): *The Civic Culture: Political Attitudes and Democracy in Five Nations*. Princeton University Press, Princeton, NJ.

AOKI, T. (2003): *Ajia Shinseiki Daiyon Kouhuku (Asias New Century, Volume 4, Happiness).* Iwanami Shoten, Tokyo.

FUKUYAMA, F. (1995): *Trust*. Free Press, New York.

LEVY, S. (Eds.) (1994): *Louis Guttman on Theory and Methodology: Selected Writings*. Dartmouth, Aldershot.

MANABE, K. (2001): *Facet Theory and Studies of Japanese Society*. Bier'sche Verlangsanstalt, Bonn.

Author Index

Baba, Y., 81
Bock, H.-H., 3
Bomhardt, C., 91
Borgelt, C., 13
Costa, I. G., 153
Douke, H., 143
Gallegos, M. T., 115
Hennig, C., 27
Johner, C., 133
Klawonn, F., 53
Knorr, T., 133
Krolak-Schwerdt, S., 41
Kruse, R., 13, 53
Manabe, K., 199
Nakai, M., 181
Nakamura, T., 143
Nakayama, A., 191
Niu, D., 123
Okada, A., 173
Ontrup, J., 163
Raabe, N., 107
Rehm, F., 53
Ritter, G., 115
Sato, Y., 61
Sato-Ilic, M., 71
Schliep, A., 153
Schmidt-Thieme, L., 133
Scholz, S., 99, 163
Schönhuth, A., 153
Siew, H.-Y., 81
Tarumi, T., 123
Wagner, R., 99, 163
Webber, O., 107
Weihs, C., 107
Wiedenbeck, M., 41
Gaul, W., 91